THE
APPLICATION OF GENETICS TO
COTTON IMPROVEMENT

THE
APPLICATION OF GENETICS
TO COTTON IMPROVEMENT

BY

SIR JOSEPH HUTCHINSON
C.M.G., Sc.D., F.R.S.

Drapers' Professor of Agriculture in the
University of Cambridge
Formerly Director, Empire Cotton Growing Corporation's
Cotton Research Station, Namulonge
Uganda

CAMBRIDGE

PUBLISHED IN ASSOCIATION WITH THE
EMPIRE COTTON GROWING CORPORATION

AT THE UNIVERSITY PRESS

1959

CAMBRIDGE UNIVERSITY PRESS
Cambridge, New York, Melbourne, Madrid, Cape Town,
Singapore, São Paulo, Delhi, Tokyo, Mexico City

Cambridge University Press
The Edinburgh Building, Cambridge CB2 8RU, UK

Published in the United States of America by Cambridge University Press, New York

www.cambridge.org
Information on this title: www.cambridge.org/9780521292559

© Cambridge University Press 1959

First published 1959
First paperback edition 2011

A catalogue record for this publication is available from the British Library

ISBN 978-0-521-05359-4 Hardback
ISBN 978-0-521-29255-9 Paperback

CONTENTS

FOREWORD

By nature I am a practical man, and I have derived great inspiration from the opportunities I have had to apply genetic knowledge to the problems of crop plant improvement. Both the genetic knowledge and the opportunities to apply it have arisen during my service with the Empire Cotton Growing Corporation, and I take this opportunity to acknowledge my indebtedness to these most enlightened employers.

No one who works on cotton and has not visited the United States Cotton Belt can consider his education complete. I am, therefore, greatly indebted to the National Cotton Council of America and to the Division of Biological Sciences of North Carolina State College, Raleigh, North Carolina, for a joint invitation to tour the United States Cotton Belt and to lecture in the Genetics Faculty at North Carolina State College in the summer of 1954. On the tour I met, in their own fields and laboratories, fellow research workers long known in correspondence. In the Genetics Faculty at Raleigh I had a greatly valued opportunity for discussion of the evolutionary background to cotton breeding problems. These made the American visit a remarkable occasion, for which I should like to express my thanks. What follows is based on the lectures delivered at North Carolina State College, with such additions as have seemed suitable in the light of the work in progress at Namulonge in the years that have elapsed since the lectures were given.

The natural starting-point for a discussion of the world's cottons is Sir George Watt's *Wild and Cultivated Cotton Plants of the World*. He published that classic monograph in 1907, and he made in his introduction an apology that is as valid today as it was in 1907. He wrote:

But as that story [of the Indian cottons] advanced from year to year I found many occasions not only to arrive at conclusions opposed to

Foreword

those held by many of my predecessors, but to modify and abandon views previously entertained, and even published, by myself.... Let me hasten, therefore, to add that I cannot hope to have even now attained finality; but shall be abundantly satisfied if I have succeded in throwing out suggestions that may tend to elucidate practical results, both commercially and botanically.

Much further progress has been made since 1907 in the study of the cottons, and it is the findings of these fifty years that will be discussed here. It is, however, one of the lasting satisfactions of the genus *Gossypium* that it continues to offer as fascinating a range of problems to the inquirer as were apparent when Sir George Watt summarised the knowledge he had then amassed.

J. B. H.

1

THE RELATIVES OF THE COTTONS

The cotton of commerce is the produce of four species of the genus *Gossypium*. The history and relationships of these four species should be considered primarily with reference to the considerable range of wild species that also belong to *Gossypium*, but some reference must also be made to the genera of the small sub-tribe of the Hibisceae (Edlin, 1935) to which *Gossypium* is related. The collection and study of the wild species of *Gossypium* has gone on for many years. Many were known to Watt as herbarium material, but the acquisition of seed and the establishment of a living collection on which cytogenetic studies could be conducted was only undertaken at a later date. The wild species are all xerophytic, and are generally to be found in rather inaccessible places. Though Kearney, working in Arizona for the United States Department of Agriculture, had some of the American wild species in cultivation in 1926, the acquisition and establishment in a living collection of all those recorded in the botanical literature was not completed until 1949. Even now, it may well be that there remain some undescribed species to be collected in the American region, and quite recently (March 1955) one new species (*G. longicalyx*, Hutchinson & Lee, 1958) was collected in central Tanganyika, and another in the Aden Protectorate.

These wild species are not only inaccessible. Most of them are rare, some very rare indeed. Their rarity, the wide distinctions between them, and the very limited variability to be found within them, have led to the conclusion that they are all of relic status, and may be without the genetic versatility necessary for further evolutionary success. As is so often the case, generalisations are misleading, and though many of the species are restricted in both numbers and distribution, there are some that must be regarded as well adapted and successful over a considerable range of desert climates.

All the wild species examined in culture save one (*G. tomentosum*) have proved to be diploid, with thirteen pairs of chromosomes. They fall into four natural cytogenetic groups, to which Beasley (1942) assigned the letters *B* to *E*. In general, crosses between species of the same group are fairly easily made. In the F_1 hybrids chromosome pairing is normal or nearly so, and at least some viable gametes are formed. Crosses between species of different groups are difficult to make and often fail completely. If F_1 hybrids are obtained, they exhibit reduced chromosome pairing, and are usually completely sterile.

The *B* group consists of two species, *G. anomalum* and *G. triphyllum*, that are native in desert regions in Africa. *G. anomalum* is widespread on the southern borders of the Sahara, occurring in a range of habitats in the five- to twenty-inch rainfall belt. Smith (1950) has shown that the perennial shrubs and trees of the dry Sudan region occur over such a belt, the ecological situation occupied changing with changing rainfall so that a roughly equivalent moisture regime is experienced wherever the plant grows. Thus, in the northern Sudan where the annual rainfall is five inches per annum, *G. anomalum* is to be found on the tops of dry rocky hills or jebels such as Gebel Merkhayat a few miles from Omdurman. Smith regards the jebel crests as the 'moist habitat' of the region, for there a plant may become established above a fissure in the rock where some moisture is accumulated. In fact, mature plants of *G. anomalum* are confined to such situations, and in the dry weather following good rains, seedlings that have failed to tap a deep source of moisture may be seen dying in considerable numbers. Further south, according to Smith, equivalent moisture conditions are to be found on the clay plains in the twenty-inch rainfall belt in Kordofan, and there *G. anomalum* is found in hollows where water accumulates after rain. That *G. anomalum* is no relic on the verge of extinction is shown by the frequency with which it has been collected in the Sudan in recent years, and by Chevalier's (1935) record that in parts of the French Sudan it is the main food of camels. The species also occurs in dry regions in the southern parts of Angola and in South-west Africa, in a form scarcely distinguishable from the northern type. There, also, the related *G. triphyllum* grows. *G. triphyllum* does not appear to be common, though it has been collected not infrequently.

The species of the *E* group are distributed over the deserts from Sind to the Sudan, and south to central Tanganyika. Only one of them, *G. somalense*, is at all common, but this has been collected over a great area from northern Kenya, through Somalia and British Somaliland to the Sudan and the north-eastern border of French Equatorial Africa. It is sufficiently variable in Somalia to have been described under five specific names. Elsewhere it is fairly uniform. Of the other species, *G. stocksii* is rare, though it has been collected in Sind, in the mountains of Dhofar, and in Somalia and British Somaliland. *G. areysianum* is confined to a very limited area in the Aden Protectorate. Two other species belong morphologically to this group, though their cytological relationships have not yet been studied. *G. longicalyx* (Hutchinson & Lee, 1958) was discovered in 1955 by Disney near Dodoma in central Tanganyika, and a plant that has been described as *Cienfugosia incana* O. Schwartz, but is certainly a *Gossypium* and morphologically one of the Stocksiana, has been collected in several places from Meifa to Ras Fartak on the coast of southern Arabia in the Aden Protectorate.*

Two Australian species, *G. sturtii* and *G. robinsonii* make up the *C* group. Both are desert plants. The former has been recorded from widely scattered localities in central and south-eastern Australia. The latter is only known from north-western Australia, and is so rare that it was not collected for more than fifty years after the original record in about 1865.

The species found on the western side of Central and South

---

* I am indebted to Miss D. Hillcoat for this information. The specimens are in the British Museum herbarium.

1-2

America are included in the *D* group. This is the most diverse, and includes the most species, of any in the genus. *G. thurberi*, which occurs in mountain valleys in Arizona and northern Mexico, is the only wide-ranging species with a large population and appreciable variability. Most of the other species are only to be found in small areas, and with very limited populations. A new species, *G. lobatum*, has recently been described from Michoacan, Mexico, by Gentry (1956) and it is not unlikely that other rare, undescribed species exist in western Mexico.

Of the cultivated species, *G. herbaceum* is the cotton of Africa and western and central Asia, and *G. arboreum* the cotton of India, South-east Asia, and the Far East. Both are diploids and belong to Beasley's *A* group. The *A* and *B* groups are related, and crosses between *G. arboreum* and *G. herbaceum* on the one hand and *G. anomalum* on the other are easily made, and give F_1 hybrids in which the chromosome pairing is good, and some viable gametes are formed (Silow, 1941). The *A* and *B* groups are the only two between which fertile hybrids have been made. The species are, in fact, so closely related that they would probably have been regarded as belonging to a single group, had it not been for the vast difference in evolutionary status between the uniform, static wild species and the highly variable, expanding crop plants.

The two cultivated species of the New World and the wild Hawaiian *G. tomentosum* are allopolyploids. There is no reason to postulate more than a single origin for allopolyploidy (see Gerstel & Sarvella, 1956). The allopolyploid complement has been shown to be made up of one chromosome set homologous with the *A* set of the Old World cottons, and one homologous with the *D* set of the wild New World species. The wild species, *G. tomentosum*, is confined to Hawaii. The two cultivated species are *G. barbadense*, the South American cotton, and *G. hirsutum*, which is of Central American origin. Both species have spread widely in the New World and beyond, and in recent times *G. hirsutum* has invaded extensive areas formerly occupied by the Old World diploid species.

Relatives of the Cottons

Three other genera of the sub-tribe, *Thespesia*, *Gossypioides*, and *Cienfugosia*, are well enough known for comparisons to be made with *Gossypium*. Of these, *Thespesia* is of particular interest in that it appears to be the one most closely related to *Gossypium*. All species examined have a chromosome number of $n = 13$, the basic number in *Gossypium*, and one species, *Thespesia lampas*, is close enough to show some genetic evidence of relationship. When used as a pollen parent on *Gossypium anomalum*, capsules are occasionally developed, and seeds are formed. The seeds, however, are invariably empty (Hutchinson, 1947).

Exell & Hillcoat (1954) have recently proposed the fragmentation of *Thespesia* into five very small genera. To do so obscures the relationships of the species, and also the existence of parallel evolutionary trends throughout the sub-tribe. The genus *Thespesia* will, therefore, be taken to include the species assigned to it by Hutchinson (1947) with the addition of Exell and Hillcoat's *Thespesiopsis mossambicensis* and Exell & Mendonca's (1954) *Thespesia acutiloba*.

Thespesia is a genus of small trees, and one woody shrub, *T. lampas*. The type species, *T. populnea*, is a circum-tropical strand plant. Related to it are a group of three species native in the islands of the Greater Antilles, and a group of three native on the coasts and hinterland of eastern Africa from Kenya to Natal. Inland in East and Central Africa and the Belgian Congo is to be found *T. garckeana*. *T. lampas* is a shrub of Indian forest lands that has been widely spread as an ornamental.

The species of *Gossypioides* have $n = 12$ chromosomes, one of them, *G. brevilanatum*, having one pair of chromosomes twice as long as the rest, suggesting derivation from $n = 13$. Two species are known. *G. brevilanatum* is very rare in relic forest vegetation in Madagascar. *G. kirkii* is common and successful in dry bush from Portuguese East Africa to Kenya. Both resemble *Gossypium* in many characters and *Gossypioides brevilanatum* is the only plant other than the cottons that bears true lint on the seeds.

5

Cienfugosia was until recently hopelessly confused with *Gossypium*. In fact it is fairly obviously less closely related to *Gossypium* than are *Thespesia* and *Gossypioides*. Chromosome numbers are known for only two species, *Cienfugosia hildebrandti n* = 11 and *C. heterophylla n* = 10. *Cienfugosia* is African and Central and South American in distribution, with a considerable range of species in each continent.

A survey of these four genera reveals a pattern of characters which forms what may be regarded as a common denominator for the sub-tribe. This common type may be described as follows:

Leaves: Entire or broadly palmately lobed.

Involucral bracts: Three, large, persistent.

Inflorescence: A many-jointed, sympodial fruiting branch.

Fruit: A dry, brittle, loculicidally dehiscent capsule, with several ovules in each loculus.

In each genus species are to be found that conform in all or nearly all respects to this pattern. Since there is no other pattern that can be regarded as common to the four genera, this may be considered to be the basic pattern from which species differentiation has proceeded.

In leaf shape, all species of *Thespesia* and *Gossypioides* have leaves of the basic pattern, either entire or more often broadly palmately lobed. In *Cienfugosia* and *Gossypium*, species are known with leaves entire or very broadly lobed at one extreme through increasing degrees of dissection to leaves that are digitately divided at the other. Variation in involucral bracts is equally great. *Gossypioides* is the only genus of the four in which all species have bracts of the basic pattern. In *Gossypium* in about half the species the involucral bracts are large and persistent, and in the rest they are reduced to a greater or lesser extent, sometimes caducous, and often no more than minute appendages. Throughout the genus they are three in number. In *Thespesia* and *Cienfugosia* the involucral bracts are, with the exception of those of *Thespesia danis*, always reduced to minute appendages and are often caducous. Frequently they are three in number, sometimes they are arranged in groups of

6

three, sometimes there are ten to twelve small appendages not arranged in groups, and in *Cienfugosia argentina* they are absent altogether.

The many-jointed sympodial fruiting branch has been regarded as the characteristic inflorescence of *Gossypium*, but on the one hand it is not universal in *Gossypium*, and on the other, in each of the other three genera there is at least one species in which it is to be found. In *Gossypium* the process of reduction of the fruiting branch is clearly seen, first in species in which there are rarely more than two nodes per branch, and then in species in which only a single node is formed, with its own subtending leaf. Then come forms in which the leaf at the node is reduced and the fruiting branch is no more than a jointed peduncle. Finally, in *G. aridum* from Mexico, the flowers are almost sessile, on short spurs which have lost all resemblance to a sympodial fruiting branch. The same sequence can be followed in each of the other three genera, but whereas in *Gossypium* the fully developed fruiting branch is common, in the other genera, most species have considerably reduced fruiting branches, or simple axillary flowers. In *Thespesia*, moreover, reduction has gone further and the terminal shoots bearing axillary flowers are in one species compressed into terminal panicles.

In fruit type, all species of *Cienfugosia*, both species of *Gossypioides*, and all species of *Gossypium* save a few of the agricultural varieties of *G. herbaceum*, have dry, brittle loculicidally dehiscent capsules. In *Thespesia* one species, *T. lampas*, has capsules of this type, whereas in the other species the capsules are more or less leathery or woody and indehiscent, or nearly so. Some agricultural varieties of *Gossypium herbaceum* also have indehiscent, rather leathery capsules. No reduction in ovule number has gone on in *Cienfugosia*, but in the other three genera species are to be found with ovule number reduced from the characteristic four to six per loculus to two to three, and in some cases to two basal ovules per loculus only.

It will be seen that the potentialities for variation are similar

7

throughout the group. There appears to be no correlated variation in the different characters, and in fact the variation between species only fits into a pattern if the basic type is regarded as primitive. In general, the basic characters are to be found in eastern and southern Africa. *Gossypium herbaceum* race *africanum* from the dry country from Angola to Moçambique and *Cienfugosia gerrardii* from the Transvaal and Swaziland are the nearest to the basic type in their respective genera. Both species of *Gossypioides*, from coastal East Africa and Madagascar, are close to the basic type. Only in *Thespesia* is the nearest species, *T. lampas* from India, to be found outside the South and East African region.

The pattern of variation demonstrated among the wild species may now be compared with that to be seen in the cultivated cottons. The course of events in the cultivated cottons can be followed with confidence, since their evolutionary history is well understood, and the identification of the primitive types is supported by historical evidence. The primitive cottons in fact conform closely to the basic pattern of the group of genera as set out above. The cultivated cottons have not followed the pattern of differentiation of the wild species. All the cottons have three large and persistent involucral bracts, their fruiting branches are in general extended, and are never reduced. With few exceptions their leaves are broadly palmately lobed, and their capsules dry, brittle and loculicidally dehiscent. The types of diversity that are to be seen in the wild species are represented in the cultivated cottons by the occurrence of more or less rare, atypical forms, Reduced fruiting branches ('cluster' and 'short branch' mutants) are known in both diploid and allopolyploid cottons. Forms with deeply divided leaves are common in some races of the diploid *Gossypium arboreum*, and occur rarely in both allopolyploid species. And in *G. herbaceum* races *persicum* and *wightianum* agricultural forms are known in Iran and Afghanistan, and in western India, in which the capsules are indehiscent. No form with reduced involucral bracts has been recorded, but deciduous

8

bract has occurred as a mutant. Thus it is evident that the genetic potentialities exist for the emergence of any of the patterns of diversity to be found in the wild species, but those patterns have not in fact developed.

The differentiation of the cottons has followed other lines. The dominant pattern of change has been the development of annual forms from primitive perennial shrubs. Annuals are not found elsewhere in the sub-tribe, but they have been developed independently in numerous separate stocks in all four species of the cultivated cottons. Capsule size, the number of seeds per loculus, and the amount of the seed-borne hairs have all been greatly increased, and the sympodial fruiting branch has in general been increased in length, rather than restricted as is so often the case in wild species. It is, of course, obvious that the characters in which the greatest change has occurred in a crop plant are those which determine its utility to man. Nevertheless, it does appear to be important to note that the pattern of change under cultivation has been radically different from the patterns that could have been anticipated from a study of the diversity among related wild species. Evidently the pattern of change is not determined by the genetic potentialities of the organism, but rather by the pattern of the selective forces to which it is subject.

2

THE ORIGIN AND SPREAD OF THE OLD WORLD COTTONS

It is convenient to distinguish between the true cottons and all other species of the genus *Gossypium*. The name 'cotton' is an agricultural and technological rather than a botanical term, and the cottons may be appropriately defined in terms of their lint, which is the character that gives them their agricultural and technological value. The seeds of all species of *Gossypium* bear hairs. The seed hairs of the wild species are short, and very firmly attached to the seed. In most wild species they are circular in cross-section and unconvoluted when mature, and remain closely appressed to the seed. In *G. tomentosum* they are flattened in cross-section and convoluted when mature, but they are still very firmly attached to the seed, and cannot be regarded as spinnable. The seed hairs of the true cottons are of two types, long, fine, flattened and convoluted hairs that are easily detached from the seed, and an undercoat of short, coarse, firmly attached hairs called fuzz. The long, fine, easily detached hairs form the lint that is spun. The short fuzz hairs are commercially useless except as a source of cellulose. In some agricultural races the fuzz hairs are absent or much reduced, giving 'naked' or 'tufted' seeds when the lint is removed. The term cotton will be restricted to those species of *Gossypium* which have seeds bearing spinnable lint hairs.

The vast majority of the true cottons are cultivated, and most of those now found wild can be identified as escapes from cultivation. Thus, although it will be shown that there are truly wild plants that bear lint, an overwhelming proportion of the cottons depend for their existence on the fact that they are useful to man.

The earlier botanists who studied the Asiatic cottons regarded the cultivated forms as having been domesticated from wild races of the same species. Numerous representatives of both species are to be found beyond the limits of current agricultural systems, and there appeared to be no difficulty in regarding them as representative of the progenitors of the cultivated cottons, or as intermediate between wild ancestral types and modern crop plants. As they were more closely studied, however, it became evident that in most cases cotton plants now to be found growing wild were confined to ecological situations that have been created by man—hedgerows, house-yards, and abandoned cultivations—and were obviously escapes from the crops formerly grown in the areas they now inhabit.

The evidence that wild cottons were descended from cultivated forms was so general that the theory was put forward (Hutchinson, Silow & Stephens, 1947) that the origin of all the diploid cottons could be accounted for by evolution under domestication, followed in some cases by escape into abandoned land or even into natural vegetation. It was argued that the diploid cottons were never truly wild, and that they had all been developed under domestication from a progenitor differing from its wild relatives in perhaps no more than a single gene mutation giving a convoluted lint hair. Since it was known from historical records that the Indus civilisation that collapsed about 2000 B.C. had played an important part in the development of the cotton crop, the suggestion was made that the original domestication of *Gossypium* might have taken place there. It was believed that the group of wild species related to the cultivated Asiatic cottons was distributed not only in South-west Africa and the region south of the Sahara, but also as far east as southern Arabia. There seemed no difficulty, therefore, in supposing that the progenitor of the cottons had been collected by travellers and introduced into the Indus Valley.

Evidence against this theory of the origin of the Asiatic cottons was soon brought forward. In the first place it was apparent that neither *G. anomalum* nor *G. triphyllum* could be regarded as representing species ancestral to the cottons since both are distinguished by advanced characters such as reduced involucral bracts and deeply divided leaves, and *G. triphyllum* also has reduced fruiting branches. The Asiatic cottons, particularly the primitive forms in both species, have the characters of the 'common type' as described above. Secondly, when *G. areysianum* was collected in the Aden Protectorate and established in culture, it turned out to be related to *G. stocksii* and *G. somalense*, and not to *G. anomalum* and *G. triphyllum* (Douwes, 1953). It thus became apparent that the species of the *G. stocksii* group (Beasley's *E* group) are to be found in the countries bordering the Arabian Sea, and *G. anomalum* and *G. triphyllum* (Beasley's *B* group), which are closely related to the cottons, must be regarded as strictly African in distribution, with their most easterly station on the Baraka river in Eritrea. The main areas of the *B* group, where one would naturally look for a related type with the more primitive characters that must have been possessed by the ancestor of the cottons, are the region immediately south of the Sahara from the Sudan to West Africa, and in South-west Africa and Angola. The chances of a traveller from the Indus Valley exploring those regions seems remote.

Further studies of the Asiatic cottons, moreover, showed that the cottons of the Indus valley do not represent the most primitive type of the Asiatic species. Hutchinson (1954) showed that in *G. arboreum*, the primitive perennial (Rozi) cotton of western India must be regarded as older than the northern cottons of the Indo-Gangetic alluvium, and Gerstel (1953) demonstrated that *G. herbaceum* is cytologically more primitive than *G. arboreum*.

A reconsideration of the problem of the origin of the Asiatic cottons was undertaken, and the conclusion was reached (Hutchinson, 1954) that a form of one of the Asiatic cottons, the South African *G. herbaceum* race *africanum*, is truly wild, and is the modern representative of the wild ancestor of all diploid cottons. Thus it appears that the essential characters of a true cotton existed before the plant was domesticated, in an old-established wild species. In *G. herbaceum* race *africanum* the seed hairs are clearly differentiated into lint and fuzz, and the lint, though wiry to the feel and firmly attached to the seed, could be ginned in the ordinary way, and could well be spun. There seems no *a priori* reason why the characters of true lint should not be established in the wild, and if they were, the plant had all the attributes necessary to attract man if he was sufficiently civilised to have learnt the art of spinning wool or flax.

As a link between the cultivated cottons and the related wild species of the *B* group, both the morphology and the geographical distribution of *africanum* are satisfactory. It has not the specialised characters of the involucral bracts and the fruiting branches that are found in *G. anomalum* and *G. triphyllum*. It resembles those species in habit, being a perennial shrub with thin and flexible branches forming a rather tangled, untidy bush. It is to be found in the dry bush region of southern Africa right across from Angola and South-west Africa to Portuguese East Africa. Its distribution therefore adjoins but scarcely overlaps the distributions of *G. anomalum* and *G. triphyllum*.

All this might have come about by the escape of a primitive cultivated cotton into natural vegetation, and the history and relationships of cultivated Asiatic cottons on the African continent must therefore be considered. It has been shown (Hutchinson, 1949) that wherever cottons of the two Asiatic species are to be found in the dry country across the savannah belt of Africa south of the Sahara, they occur in cultivation, in houseyards, on field margins and hedgerows, or on abandoned land that was formerly cultivated. They are associates of man and his agriculture, and are not truly wild. On the coast of East Africa a form of *G. arboreum* closely resembling the perennial Rozi cotton long cultivated in western India has occasionally been collected. This form spreads as far as Madagascar, where it is established in natural vegetation in the drier parts of the island (Hutchinson, 1954). This cotton also must be regarded as an associate of man, since there is evidence that its distribution is related to its use in a spinning industry.

An account of the growth and use of cotton on the East African coast was given by Duarte Barbosa (1866) who visited Sofala, in what is now Portuguese East Africa, about the beginning of the sixteenth century. He recorded that there was a regular trade in cotton goods from Cambay in western India by way of southern Arabia and East African ports to Sofala in exchange for gold from the kingdom of the Monomotapa (the present Zimbabwe). Moreover, he stated that the 'Moors' (Moslem Arabs) had recently begun to grow cotton and spin

13

and weave it locally. Cambay, whence the cotton goods came, is the home of the Rozi form of *G. arboreum* from which the *arboreum* cottons of East Africa and Madagascar are almost indistinguishable. It can scarcely be doubted that in setting up their own cotton industry, the Arab traders would use the cotton of the country whence they had previously brought their trade goods. Thus the occurrence of *G. arboreum* all along the Indian Ocean trade route may be ascribed to its spread from the trading centres of western India.

There is no reason to associate *G. herbaceum* race *africanum* with this trade route. Not only is it spread right across the continent, far beyond the areas affected by the gold trade, but it has never been recorded from the trading ports of the East African coast. It must therefore be regarded as truly wild in southern Africa. In no other area is *G. herbaceum* to be found beyond the limits of agricultural land, or land on which the vegetation has been disturbed by man, and *africanum* appears therefore to represent the wild form from which the cultivated races of *G. herbaceum* were derived.

The cottons were certainly not originally domesticated in southern Africa. There is no evidence of spinning and weaving in that area before the time of which Barbosa wrote, and the industry he saw was short lived. The domestication of a new textile plant, and its evolution into the range and diversity now to be found in the cottons, could only have been achieved by a large, long-persisting civilisation or sequence of civilisations, in which strong, enterprising and versatile textile crafts were developed.

By whom, and at what date, wild cottons were first brought into cultivation must remain a matter for speculation, but it is probably significant that *africanum* is indigenous in the area inland from Sofala that has been from time immemorial a source of gold. Trade round the Indian Ocean coastline between western India (Cambay) and Sofala was established long before Barbosa's time. It was largely organised by the peoples of southern Arabia. The most primitive of the cultivated forms

Spread of Old World Cottons

of *G. herbaceum* are those of race *acerifolium* to be found in southern Arabia. These, with their relatives along the Arabian Sea coast of Iran and Baluchistan, lie along the trade route by which their ancestors were presumably brought from southern Africa. The extension of the race into northern Africa has been shown (Hutchinson, 1949) to be a later development associated with the westward spread of Moslem power. It is a fact that has escaped comment, but now appears significant, that the old records as far back as the accounts of Theophrastus in 350 B.C. (see Watt, 1907) link Arabia with India in discussing the cultivation and use of cotton. Doubtless by the beginning of the Christian era India, with its vast population and more favourable climate, had far outstripped Arabia, but it may well be that in the early days of human civilisation the rainfall in Arabia was greater, and the climate more favourable to cotton than it is at present. Be that as it may, the evidence of botanical relationships and the geography of the ancient trade routes agree in indicating southern Arabia as the probable locality of the first domestication of cotton.

The sequence of development of the annual races of *G. herbaceum* presents no difficulty, once the establishment of *acerifolium* on the Arabian and Iranian coasts is accepted. Spread northward was only possible when an annual type, able to crop and finish before the onset of cold winters, had been established. Thus arose race *persicum*, and from it, by further development along the same lines, the very early cropping race *kuljianum*, adapted to the short, hot summers of central Asia.

No date can be fixed for the establishment of the annual habit, but Marco Polo (A.D. 1290) reported that cotton was grown in what is now Chinese Turkestan at the time of his visit, and no perennial cotton would stand the winters of central Asia. By contrast, Marco Polo recorded that the cottons of India at that time were perennials that lived as long as twenty years.

The development of the annual Indian race *wightianum* can be dated fairly well. Dr Hove collected in Gujerat, in A.D. 1787,

15

the perennial form of *G. arboreum* known as Rozi, and also the annual *G. herbaceum* race *wightianum*. Watt (1907) comments 'Dr Hove's specimens are in the British Museum, and it has to be admitted that they could not be separated botanically from any corresponding set of more recent date'. On Hove's *wightianum* material Watt says, 'strangely enough, his remarks regarding it almost involve the belief that it was then a new crop...'. With the knowledge now available on the origin of the annual habit it no longer appears strange, and it may be postulated with some confidence that *G. herbaceum* race *wightianum* was developed in western India following the introduction of annual, open bolled forms of the species from Persia in the early eighteenth century. The closed boll character found in many *persicum* types and in some forms of *wightianum* from western India are not characteristic of the species as a whole, or even of the early *persicum–wightianum* line. They are best regarded as a recent specialisation in *persicum* that spread to western India after the first establishment of *wightianum*.

From Hove's account it may be seen that the perennial Rozi, which is a primitive form of *G. arboreum*, was grown in western India before *G. herbaceum* race *wightianum*. *G. arboreum* is in fact very ancient in India. Marco Polo's (A.D. 1290) perennial cotton of Gujerat, which grew and cropped for as much as twenty years, was almost certainly a form similar to Rozi. Moreover, the oldest known specimens of Asiatic cotton, the fragments of fabric recovered from Mohenjo Daro in Sind (*c.* 3000 B.C.) and examined by Gulati and Turner (1928), appear to have belonged to *G. arboreum*, and not to *G. herbaceum*.

Gerstel (1953) has shown that *G. arboreum* differs cytologically from *G. herbaceum* by one translocation in the chromosome complement, and also that the arrangement in *G. herbaceum* is the same as in *G. anomalum*, and so may be regarded as the more primitive condition. Genetically, *G. arboreum* and *G. herbaceum* are closely related. Crossed together they give a vigorous F_1 which is highly female fertile.

Spread of Old World Cottons

The existence of the translocation results in partial male sterility. That they are indeed specifically distinct is shown by three independent lines of evidence: (1) The species integrity is maintained when cultivated forms are grown mixed in commercial crops. Mixtures of *G. arboreum* and *G. herbaceum* have been grown commercially in western India and in parts of Madras for many years without breakdown of the species distinction (Hutchinson, Silow & Stephens, 1947; Mudaliar & Balasubramanyan, 1951), although vigorous F_1 hybrids are to be found not infrequently in the field. (2) Before the distinction between *G. arboreum* and *G. herbaceum* was worked out genetically, comprehensive and long-continued efforts were made in South India to breed commercially acceptable cottons from *G. arboreum* × *G. herbaceum* hybrids. These consistently failed to give material of agricultural value, and it was remarked that 'the better the single plant selection in any generation, the worse the segregates that appeared in its progeny'. Comparable material in the same breeding scheme from crosses then believed also to be inter-specific, but between races subsequently shown to belong to *G. arboreum*, gave commercially acceptable derivatives without difficulty. (3) In crosses made for the genetic analysis of the species difference, there arose a wide range of unbalanced and unthrifty types in segregating generations.

It seems probable that *G. arboreum* arose from *G. herbaceum* in the course of development of the Asiatic cotton crop. It has been shown (Hutchinson, 1954) that the types of *G. arboreum* now wild in India, the Sudan and West Africa, and South-east Asia may be regarded with some confidence as escapes from cultivation, and the East African coastal series can be related to the spread of cotton textiles and the cotton crop from the ancient textile-exporting civilisation of Cambay in western India. Madagascar types are indistinguishable from the perennial forms of western India (Watt, 1907). Since throughout the east and south of Madagascar, where cottons occur, plants introduced by man are to be found in process of acclimatisation and established in natural plant associations (Renaud Paulian, in correspondence), the *arboreum* cottons of the island may be regarded with some confidence as the end point of the Cambay–East African coastal sequence.

The extensive differentiation of agricultural races in *G. arboreum* has been elucidated by Silow (1944), and Hutchinson, Silow & Stephens (1947) and Hutchinson (1954) have amplified the earlier account. *G. arboreum* race *indicum* in its more primitive perennial forms in western India (Rozi) is morphologically so similar to *G. herbaceum* races *acerifolium* and *wightianum* that there has been much confusion between the species. Genetically, Silow (1944) has shown that race *indicum* is closer to *G. herbaceum* than is any other race of *G. arboreum*. It may then be supposed that the development of *G. arboreum* took place in the home of the primitive perennial forms of its race *indicum*, which is in western India. Thus the association of the early development of the cotton crop with the trade routes of the Indian ocean coasts is carried a step further. It may well be that Cambay, the home port of the cotton trade in Barbosa's time, had been the centre of commerce in cotton since the time that *G. arboreum* first became established.

In Peninsular India, in response to the demand for fine hand spun and hand woven fabrics for the upper classes of the Hindu civilisation, some of the best quality strains found in the Asiatic cottons were developed. In more recent times (Ramanatha Ayyar, 1938) stocks were selected for the annual habit, so that there are now only a very few small areas in which perennials are still grown.

Hutchinson (1954) has pointed out that the races of *G. arboreum* other than race *indicum* form a natural group that is rather markedly contrasted with race *indicum*. Their centre appears to be in the region of the great Indo-Gangetic alluvium of northern India, and it is suggested that they are the cottons developed and spread by the Indus civilisation of which Mohenjo Daro (see Gulati & Turner, 1928) was one of the great cities. These northern *arboreums* are pyramidal shrubs with long flexible branches, and are highly variable in leaf shape, flower colour, and anthocyanin pigmentation. By contrast, the *indicum* cottons are rounded bushes when well grown, with broad leaves, little anthocyanin pigmentation, and yellow flowers. Moreover, in general the northern cottons are short and coarse with a high ginning out-turn, whereas the *indicums* are generally low in ginning out-turn, and are comparatively fine in staple.

The primitive forms of the northern *arboreums* are perennial. They are very widely spread, being found as associates of man in field margins, gardens and house compounds, all through northern India, eastwards through Burma and Indonesia to the Philippines, and westward across southern Arabia, the northern Sudan, and the southern borders of the Sahara to West Africa. They provided the stock for the first African cotton crops, at Meroe in the northern Sudan (Hutchinson, 1949). In the 2000 or so years since they reached the Nile they have developed a sufficiently distinct type to be recognisable as a geographical race—race *soudanense*—but in Africa they were never cultivated so intensively as to give rise to an annual form.

In India and the Far East very important commercial crops were built upon the northern *arboreums*, and annual races were developed. In India the coarse, short, high ginning *bengalense* eliminated its perennial ancestors from the crop lands, and invaded extensively the area of the finer but lower ginning and poorer cropping *indicums*. In Burma and Indonesia, the commercial crop came to depend primarily on annual types of race *burmanicum*, but perennials have maintained a place. In China, Manchuria, Korea and Japan, only early annual cottons could survive, and the spread of cotton into that region followed the development of the early annual habit now characteristic of race *sinense*.

The sequence of development of the Asiatic cottons was thus as follows: from the wild *G. herbaceum* race *africanum* the perennial, primitive cultivated race *acerifolium* was developed.

Northward spread followed the development of the annual habit, and led to the establishment of races *persicum* and *kuljianum*. From the primitive perennial *G. herbaceum* which spread into India, arose the earliest form of *G. arboreum*. From western India *G. arboreum* spread into the alluvial areas of what is now West Pakistan, giving rise to the perennials of the northern form of *G. arboreum*. These three perennials spread over the areas in the Old World that were suitable for the growth of perennial cottons. The success of cotton as a textile material was such that cottons were required for areas where perennial growth was impossible. The need was first met in India by importing annual *herbaceum*, and this gave rise to race *wightianum*. With the demands of a more intensive agriculture in monsoon regions, there followed the development of annuals from perennial *G. arboreum* throughout the commercial producing areas, and in each of the three major groups the perennials are now of relic status, having been supplanted by their own annual derivatives.

3

THE ORIGIN AND SPREAD OF
THE NEW WORLD COTTONS

Three of the species of *Gossypium* are allopolyploids, with twenty-six pairs of chromosomes. All three bear convoluted hairs on their seeds. Two of them include the cultivated cottons that were originally confined to the New World, and certain relatives which, though now found growing wild, can be shown in the great majority of cases to have been derived from cultivated forms. The third, *G. tomentosum*, is a wild species endemic in Hawaii.

The relationship between the two American species, *G.*

barbadense and *G. hirsutum*, and the diploids has been the subject of extensive cytogenetic studies, begun by Skovsted and carried on by Beasley, Stephens and Gerstel. Skovsted (1937) showed that basically the chromosome complement of the allopolyploids was made up of one set of thirteen homologous with that of the Asiatic cottons and one set homologous with that of the New World wild species. Beasley (1942) confirmed Skovsted's deduction and made the first synthetic allopolyploid by doubling a sterile hybrid between *G. arboreum* and *G. thurberi*. On the nomenclature he proposed, the allopolyploids have a chromosome complement symbolised as $2(AD)$, and his synthetic polyploid, which was $2A_2 2D_1$, proved to be cytologically homologous with the New World cottons. It differed considerably from the New World cottons, however, and was not fully fertile. Stephens (Hutchinson, Silow & Stephens, 1947), who was able to study a wider range of the wild diploid New World species than was available to Beasley, concluded that the wild Peruvian *G. raimondii* is more closely related to the New World cottons than is *G. thurberi*, or any of the other species of the *D* group. Gerstel (1956) has recently added to the evidence in support of Stephens's view from his studies of segregation in hexaploids involving *G. raimondii* × *G. hirsutum* and *G. thurberi* × *G. hirsutum*.

Until recently, since the New World diploid species and the more primitive of the allopolyploids are to be found on the Pacific slopes of the American continent, it was assumed that *G. arboreum*, which spreads as far as the Pacific coasts of Asia, was the closest relative of the Asiatic progenitor of the polyploids. Gerstel (1953), however, has shown that the chromosome configuration of the New World species is closer to that of *G. herbaceum* than it is to that of *G. arboreum*. The *A* set of the chromosome complement of the New World cottons differs from that of *G. herbaceum* by two translocations. From that of *G. arboreum* it differs by three translocations. So small a cytological difference may well be regarded as unimportant, since, in many genera, chromosome interchanges are common within the limits of a single species. Chromosome behaviour in *Gossypium* is so regular, however, that such a difference cannot be disregarded. Taken together with the evidence reported above that *G. herbaceum* is the more primitive of the two Asiatic species, having the same chromosome order as the wild *G. anomalum*, it indicates that *G. herbaceum* is more likely than *G. arboreum* to be the nearest relative of the *A* bearing parent of the allopolyploids.

The central problem of any discussion of the origin of the New World cottons is to show how the parent forms came together so that hybridisation could take place. *G. herbaceum*, which originated in southern Africa, never extended eastwards even as far as the coasts of the Bay of Bengal, while *G. raimondii*, on the other hand, is confined to a few river valleys in dry country on the Pacific coast in north Peru. Three hypotheses have been put forward to account for the meeting of the parent species and the origin of the first polyploid. The first was Harland's (1939) theory of a trans-Pacific land bridge. Having noted the occurrence of *G. tomentosum* in Hawaii, *G. barbadense* race *darwinii* in the Galapagos Islands, and *G. hirsutum* race *punctatum* (the form then known as *G. taitense*) in Polynesia, he quoted Schuchert, who postulated the existence of a trans-Pacific land bridge in late Cretaceous or Tertiary times, and supposed that polyploidy had arisen in the Pacific region when the parental types might have met overland. There seems little support for the view that such a land bridge ever existed, and in addition the evidence indicates that the distribution of the allopolyploid cottons in the Pacific region is of recent origin. *G. barbadense* and its var. *darwinii* in the Galapagos Islands appear to be fairly recent immigrants from continental South America. The *G. barbadense* types found in other islands in the Pacific were undoubtedly carried there after the exploration of the Pacific by European nations. The primitive forms of *G. hirsutum* in the Marquesas Islands, Tahiti and Fiji have their close relatives in southern Mexico, and their dependence on man is indicated by their occurrence in or near villages or old village sites (Hutchinson, Silow & Stephens, 1947). Only *G. tomentosum* is truly Polynesian in origin, and the occurrence of an endemic species in Hawaii can hardly be regarded as evidence in support of the Pacific land bridge theory.

Stebbins (1947) has suggested a modification of Harland's theory. He considers that allopolyploidy may have taken place in North America following the spread thither of Asiatic cottons via northern Asia and Alaska in early Tertiary times. Stebbins states that 'the subtropical woody flora characteristic of the Eocene deposits of the United States contains a mixture of Asiatic and New World elements, so that it is highly probable that a similar mixture was found among the herbaceous species, including *Gossypium*'. There were no herbaceous species of *Gossypium* at that time, and there is no member of the genus that would survive in a range of vegetation types of which a subtropical woodland was a dominant member. *Gossypium* is to be found in open, dry vegetation, and moreover all save the most recent agricultural forms are perennial shrubs, which are susceptible to frost. The most mesophytic of the wild species is *G. herbaceum* race *africanum*. Even this is confined to semi-arid tropical bush country, where the small trees and short, tufted grasses leave ample space for the development of light-loving seedlings. The annual habit is a direct consequence of human selection for agricultural suitability (Hutchinson, Silow & Stephens, 1947), and it is only among the annual cottons that spread into regions with cold winters is possible. A southerly route by way of Antarctica in Tertiary times or earlier, when conditions there are believed to have favoured plant dispersal between the countries of the southern hemisphere, is unacceptable for the same reasons. Natural spread of wild species of *Gossypium* can only be regarded as possible if it is shown that the line of spread was across a frost-free, arid or semi-arid area.

Gerstel (1953) has pointed out that since it now appears that the African *G. herbaceum* is nearer than *G. arboreum* to the ancestral line of the New World allopolyploids, the possibility that the parents of the allopolyploids came together across the Atlantic, and not across the Pacific, must be taken into consideration. The chief obstacle to a trans-Atlantic meeting of the diploids lies in the distribution of the putative diploid parents. The American diploids are all to be found on the western side of the American continent. In southern Africa, though *G. herbaceum* race *africanum* has been recorded in southern Angola and South-west Africa in the territory of *G. triphyllum*, its main, and doubtless original, area is on the eastern side of the continent in Southern Rhodesia, Portuguese East Africa, and the low veld of the Transvaal.

There are in Brazil semi-arid areas that would be ecologically suitable for the establishment and development of true cottons. Cottons of *G. hirsutum* race *marie-galante* are, in fact, well established there, but all the evidence (Hutchinson, 1951) indicates that they arrived there by migration from Central America, and were not evolved in that region. Thus, in so far as present distributions are a guide to past contacts, the formation of the allopolyploids following an Atlantic crossing must be considered most improbable.

An alternative to natural spread by a wild diploid across the Pacific region is that the allopolyploids arose since the dawn of civilisation, and that the Asiatic parent was a diploid cotton that was introduced by man into the American continent. This suggestion was made by Hutchinson, Silow & Stephens (1947), and it has given rise to considerable controversy. Many students of evolution, taxonomic botany and plant geography consider that the rates of evolutionary development postulated are too high to be acceptable, and that the evidence for a recent origin following Pacific crossings in pre-Columbian times is unconvincing. They believe that other opportunities of contact between the putative parents of the allopolyploids may have existed in the distant past. On the other hand, those botanists, ethnologists and geographers who have other reasons for thinking that there were cultural contacts between the Old and New World early in the history of civilisation, have noted with interest the theory that one of the diploid ancestors of the New World cottons was carried across the Pacific by man.

The chances of obtaining direct and conclusive evidence on the age of any plant group are very remote, and they are only slightly better among crop plants than among other sections of the world's flora. Harland (1950–1) indeed concluded: 'We can only assess probabilities.... The two schools [ancient *v.* recent origin] will continue to engage in controversy—and we shall get no further.' In so far as theories of ancient origin carry the corollary that the relevant events occurred so long ago that no evidence of them now remains, this counsel of despair may be valid. If the New World cottons are of recent origin, however, further progress is possible since a reconstruction of their history can be made, and tested by the prediction and verification of its consequences. It is intended here to examine critically the information that has accumulated on the early forms of the New World cottons, and to assess the value of the recent origin theory in elucidating their history and development.

On the theory that the Asiatic parent was brought in contact with the New World parent by trans-Pacific migrants, it was suggested that the earliest users of cotton in the New World must have settled in the area of North Peru now occupied by *G. raimondii*, which is the nearest New World diploid relative of the allopolyploid cottons. Taking this suggestion into account, an archaeological expedition from the American Museum of Natural History in 1946 chose for excavation an ancient Peruvian habitation site, the Huaca Prieta, at the mouth of the Chicama valley in North Peru (Bird & Mahler, 1951–2). There they found the oldest cottons so far discovered in the New World. The results of the expedition have been the subject of preliminary reports by Bird (1948*a*, 1948*b*), Whitaker & Bird (1949), Bird & Mahler (1951–2) and Bird (in Johnson, 1951). The Huaca Prieta is a large mound made up of the accumulated refuse of human habitations. The colony that gave rise to it was founded by 'a group of people totally unlike the earlier inhabitants of the same region' (Bird & Mahler, 1951–2), and it was built up and abandoned before pottery was used in those parts, though the introduction of the ceramic art at a later date was traced in a neighbouring site. The people of the Huaca Prieta were primarily fisher folk, but they practised a primitive agriculture from the earliest times. Their crop plants included 'squash, chili peppers, beans of possibly four varieties, achira (Canna), bottle gourds, and cotton' (Bird, 1948 b). Whitaker & Bird (1949) suggested that the beans were probably all *Canavalia*. No trace of maize was found in preceramic layers. Estimates of the age of successive levels were made by the radiocarbon dating technique (Johnson, 1951). Material from the bottom of the mound was dated at about 2400 B.C., and layers of a neighbouring site in which the first pottery was found at about 1225 B.C.

The fabrics found in the Huaca Prieta have been discussed by Bird & Mahler (1951–2). They concluded that 'the textile craft of Peru was based primarily on the use of cotton'. A bast fibre was used, but was subsidiary, and wool did not appear until very much later (not before 1000 B.C.). Over three-quarters of the articles recovered were made by twining, and less than 5 per cent by weaving. Fish nets made up about one in twelve of the items listed. Evidently the textile art of the Huaca Prieta was extremely primitive in comparison with that of Mohenjo Daro in Sind (Gulati & Turner, 1928) which was roughly contemporary with it.

The dependence of the people on fishing is an important consideration. According to Bird & Mahler (1951–2) the Huaca Prieta and other contemporary sites appear to have been chosen initially, and to have been occupied so long, because they were in the few places 'that offered the desirable features of land which could be cultivated without irrigation, and protected sections of the beach where net fishing was possible'. Considering the importance of net fishing and the primitive nature of the fabrics other than nets, it seems fair to conclude that the primary use of cotton was for fishing nets.

Spread of New World Cottons

The Huaca Prieta material is of considerable significance for evolutionary studies of cotton. The cotton used was primitive in all fibre characters. The technique of making twine was fairly good, presumably in connection with the manufacture of fishing equipment. But the mastery of the textile art was very poor, and textile-making tools were extremely primitive or absent. Cotton material from a series of layers from the earliest up to those in which pottery and maize were also found was made available for examination through the courtesy of Junius B. Bird of the American Museum of Natural History. There was nothing in the material to suggest that more than one species of cotton had been used. Many of the seeds, especially some tufted seeds from layer H.P. 3, resembled closely the seeds of modern types of *G. barbadense*, and the material can be assigned to this species with a fair degree of confidence. H.P. 3, Q was dated (Johnson, 1951) *c.* 2400 B.C. There was nothing in the material to suggest the presence of an Asiatic cotton.

Processed lint was observed in three samples. The oldest was a small piece of a fine, well-made twofold string in H.P. 3, Q. 1 (*c.* 2400 B.C.). The component yarn as well as the finished string was very even, indicating considerable technical skill in production. In the sample from H.P. 3, I. 1 (*c.* 1700 B.C.) was a small quantity of cream lint spun into a very characteristic roving that was immediately recognisable as a lamp wick. A small tangled knot of yarn was found in H.P. 3, N, a layer between Q (*c.* 2400 B.C.) and M (*c.* 2300 B.C.). It was rather coarse and irregular, and not comparable in technique with that used to make the string. A portion of it was submitted to the British Cotton Industry Research Association's Shirley Institute, for an examination of the measurable characters of the lint from which it was made. The director of the Shirley Institute wrote:

The material is very tender and undoubtedly was broken in testing. After gently 'hand-pulling' the cotton we estimate the effective length is probably about $\frac{3}{4}$ inch. We have made an immaturity count and,

27

from the figures of this and the fibre weight test, estimated in the usual way the corresponding figure for standard fibre weight. The results should, however, be accepted with caution because age may have affected the swelling characteristics to some extent. Nevertheless it appears safe to say that the cotton is short, coarse and immature.

Data for fibre weight and fineness were as follows:

Fibre weight per cm.	186×10^{-5} mg. per cm.
Immaturity count (normal %–dead %)	9–44
Standard fibre weight	355×10^{-5} mg. per cm.

The sample from the slightly older H.P. 3, Q. 1 layer included both yellow and brown lint so it is evident that the brown colour can survive under these circumstances and it may be concluded that the original colour of the 'yellow' lint was white or light brown. An examination by J. P. Evenson of a sample from a layer of the same deposit, dated about 1500 or 1600 B.C., gave comparable figures. Length was estimated by sorting out the longest fibres and measuring them under suitable magnification. The mean length of the hairs in the sample was 26 mm., which is equivalent to an effective length of about $\frac{7}{8}$ inch. Hair weight and standard hair weight were estimated at 186×10^{-5} mg. per cm. and 309×10^{-5} mg. per cm. respectively. Thus the two samples were very similar, though separated by a time interval of the order of 800 years.

G. barbadense is at present widely distributed in South and Central America and the West Indies, but the more primitive forms, and the few representatives of the species that are now to be found beyond the influence of cultivation, are confined to western South America. The primitive *G. barbadense* collections that have been studied in culture include five types from the Galapagos Islands supplied by officers of the United States Department of Agriculture and a series obtained by E. K. Balls and others in Bolivia, Peru and Ecuador. These have been discussed elsewhere (Hutchinson, Silow & Stephens, 1947), but their importance in a study of the origin of the allopolyploid cottons justifies a more detailed report of the records of their natural status.

Spread of New World Cottons

From a study of the five Galapagos types it was concluded that there is no line of demarcation between var. *darwinii* and the typical form of *G. barbadense*. After examining the much larger collection of dried specimens in the Gray Herbarium, R. A. Silow wrote (MS. notes): 'When I first went through these sheets, I tried to separate them into narrow leaved typical *darwinii* and broad leaved *darwinii*, and typical *barbadense*. I concluded that the material on each sheet was not sufficiently complete to do this (plants obviously growing in a very dry site), but further that such a separation was probably not justifiable.' All the Galapagos material is, of course, wild in natural vegetation. The forms that have been examined in culture vary from those that have the small seeds and hard seed coats characteristic of the wild cottons, and brown, scanty* lint that is firmly attached to the seed, to those that have seeds more like *barbadense* proper, and more copious, lighter coloured lint. In the forms examined the lint had a maximum length of 27–30 mm., which is equivalent to an effective length of one inch.

The mainland types studied were obtained from the modern commercial crop (which is mostly Tanguis), from country markets where seed cotton was exposed for sale, from small cultivations, from field margins, and in the wild. E. K. Balls supplied field notes on his collections from which the following may be quoted:

Rio Abajo, Bolivia

Growing wild along edges of cultivated ground, and in spare spaces in woods, not cult., but cotton collected and used in making clothing.

Limatambo, Apurimac, Peru

High dry bank on edge of maize field. Not cult., but volunteer seedlings encouraged by natives who use the cotton.

Guayaquil, Ecuador

Common in woods and thickets, and apparently more or less cultivated also.

* The ginning out-turn of one of the *darwinii* types was about 12 per cent as against 25–35 per cent in the commercial forms of *G. barbadense*.

29

Genetics and Cotton Improvement

From these wild or semi-cultivated types the transition is easy to the Tanguis of the commercial crop on the one hand and through intermediates such as Svenson's (quoted by Stebbins, 1947) from Ecuador, and one collected by Boza at Tumbes in Northern Peru, to the var. *darwinii* of the Galapagos Islands, on the other. On the mainland where cotton is in regular use the line of demarcation between wild and cultivated vanishes, and the crop of naturally growing plants is harvested for domestic use.

Data on fibre characters are available from a number of these primitive South American types, and are set out in Table 1 below.

Table 1. *Fibre characters of* G. barbadense *cottons from western South America*

Source	Ancient Peru		Ecuador	Bolivia
	Huaca Prieta		wild	Rio Abajo wild
	H.P. 3 N.*	G. 3†		Light
Lint colour	Cream	Cream	White	brown
Effective length (1/32 inch)	24?	28?	36	32
Fibre weight per cm. ($\times 10^{-5}$)	186	186	322	201
Immaturity count	9–44	—	55–16	26–29
Standard fibre weight ($\times 10^{-5}$)	355	309	360	293

Source	Peru				
	Pais	Semi Aspero	Pardo	Pardo	Pardo
Lint colour	White	White	Cream	Cream	Cream
Effective length (1/32 inch.)	43	50	49	52	51
Fibre weight per cm. ($\times 10^{-5}$)	183	175	149	142	127
Immaturity count	54–16	59–16	52–17	58–14	49–21
Standard fibre weight ($\times 10^{-5}$)	206	191	170	154	151

* Examined by Shirley Institute. † Examined by J. P. Evenson.

The Huaca Prieta data are included for comparison. The modern samples are ranked in order from the coarsest to the finest, on standard hair weight. Allowing for the uncertainty inevitably associated with measurements of ancient material, it appears reasonable to regard the Huaca Prieta samples as approximating to the shortest, coarsest and most immature *barbadense* cottons of the present day. This conclusion is important from two points of view. First, apart from rare types with highly coloured lint, the combination of very low maturity and high hair weight is unknown outside *G. barbadense*, and the belief that the Huaca Prieta cotton was a form of *G. barbadense* is thereby strengthened. Secondly, the quality of the cotton in the earliest known textiles was no better than that of cottons now regarded as wild in the same region. This is significant in view of Stebbins's (1947) argument that the fibre of the cottons now found wild in natural plant associations in Ecuador and north Peru is worthless, and that since it is worthless, and is not used by the local people, the plants that bear it are truly wild, and are not cultivation escapes. Not only is there no real line of distinction between those that are used by the local people and those that are not (see notes by E. K. Balls quoted above), but the modern wild-growing material that has been examined is of a quality equal to or better than that in the oldest textiles known from the area.

These primitive cottons of western South America may be regarded as representative of the source material from which the *barbadense* cottons of other areas arose. Locally they gave rise to the extensive modern Tanguis crop in Peru. Even this is fairly primitive by the standard of most cotton-growing countries, since Tanguis is a perennial, cropping for several seasons before being replanted. From western South America, *G. barbadense* spread eastwards through the forest regions of Brazil, Venezuela and the Guianas, giving rise to the large-leaved, large-bolled 'kidney'-seeded race *brasiliense*. North-eastwards through Colombia, earlier-fruiting, free-seeded types spread to the coasts and islands of the Spanish Main and the Guianas. There they met *brasiliense* spreading north, and there is consequently almost as great a range of perennial *G. barbadense* in the West Indies as in Peru.

It was from this range of material that seed was sent to South Carolina, and gave the stock out of which the Sea Island cottons were selected. Hutchinson & Manning (1945) have suggested that along the Colombia–Spanish Main–Greater Antilles line

31

of spread some of the earlier-fruiting, finer-quality Peruvian lines reached Jamaica, and later provided the earliness and good quality without which the Sea Islands could never have been developed.

Perennial forms of *G. barbadense* were introduced into many parts of Africa. There is no record of the sources of the introductions, but since all the recognisable South American and West Indian types are to be found in Africa (particularly in the forest regions of West Africa), it is clear that seed must have been brought in from many places in eastern South America and the Caribbean region. The *barbadense* cottons established in West Africa spread across by the trade and slave routes to the Sudan, and a few specimens were taken thence by officials of the Egyptian government and grown in their gardens in Egypt. In 1820 Jumel recognised the potentialities of one of these perennial cottons, and was successful in establishing commercial cotton growing in Egypt. The development of the annual Egyptian cottons, which will be discussed later, followed in the middle of the nineteenth century, and by the beginning of the present century the very distinct race of Egyptian cotton was fully established both in the agriculture of the Nile valley and in the cotton markets of the world.

One of the most interesting features of the history of *G. barbadense* is the extent to which its evolutionary progress can be dated. If the Huaca Prieta cotton be accepted as representing one of the earliest cultivated *barbadense* cottons, the time scale of Table 2 may be set out:

Table 2.

c. 2500 B.C.	Huaca Prieta, coarse, short, presumably perennial.
A.D. 1785	Introduction into Carolinas, annual habit established.
A.D. 1820	Jumel's cotton brought into commercial cultivation.
c. A.D. 1850	Beginning of development of annual cottons in Egypt.
A.D. 1907	The Sakel variety bred in Egypt.
c. A.D. 1935	Superfine V 46 (Hutchinson & Manning, 1945) put into commercial cultivation in St Vincent.

In the 4000 years or so between the Huaca Prieta culture and the introduction of *barbadense* into Carolina, it may reasonably

32

be supposed that differentiation resulted in the establishment of race *brasiliense* in the forest region, and the development of considerable diversity throughout the species in habit and in quality characteristics. In the next 150 years the annual habit was established in the Sea Islands and in Egypt, and great strides were made in productivity. In the Egyptian cottons, which are the most recent, the development of early, heavy cropping strains is still going on. Over the same period of a century and a half there was developed, first in the Sea Islands and then in the Egyptian cottons, a length of staple and a fibre fineness beyond anything previously known, culminating in the Sea Island Superfine V 46 with an effective length of two inches, and a standard hair weight of 114×10^{-5} mg. per cm. The whole sequence provides a very well-dated example of the way in which the rate of evolutionary change can increase in response to the opportunities offered by new environmental circumstances.

The nature and distribution of diversity in *G. hirsutum* has been greatly illuminated by the Central American collections made by Richmond, Stephens, and Ware and Manning. These were all grown, together with an extensive collection of *G. hirsutum* from commercial crops in other parts of the range of the species, at Shambat in the Sudan, and the results of a comparative study published (Hutchinson, 1951). The conclusion (Mauer, 1930) that the centre of the species *G. hirsutum* is in southern Mexico and Guatemala was confirmed, and evidence of extensive differentiation within the species was presented.

Seven races were identified in the Central American material, each with its own geographical area. Four of these, *palmeri*, *morilli*, *richmondi* and *yucatanense*, are not known outside the Central American region. The first three are perennial shrubs, commensal with man in house-yards and hedgerows, and evidently the relics of primitive cultivations. The fourth, *yucatanense*, was shown by Stephens, who collected it, to be a race developed from the more widespread race *punctatum* for the specialised coastal sand dune habitat of northern Yucatan. Though *yucatanense* is wild, the parental race is cultivated, commensal, or only secondarily wild.

The three races *punctatum, marie-galante* and *latifolium* are well known, as they have spread widely outside Central America. *Marie-galante* is represented in Central America by types that are smaller and poorer in lint quality than the peripheral forms, which are the large tree cottons to be found all along the Spanish Main, south-east to north Brazil, and north-east into the West Indies. These Caribbean tree cottons vary greatly in quality, and there is reason to believe that they have gained in staple and fineness by introgression from some of the finer components of *G. barbadense*. The fine Colombian *marie-galantes* and the fine 'French' or 'Small Seeded' cotton (Hutchinson & Stephens, 1944) both had their origin on the line of spread postulated by Hutchinson & Manning (1945) for the fine *barbadenses* that gave rise to the Sea Islands.

The *punctatums*, which are the cottons of eastern Mexico and Honduras, spread through the coasts and islands of the Gulf of Mexico as far as the Bahamas. The *marie-galantes* spreading north through the West Indies and the *punctatums* spreading east through the islands of the Gulf of Mexico overlap in the Greater Antilles. The *punctatums* never achieved in the New World the status as crop plants attained by the *marie-galantes*, but they were carried much farther in the Old World. The *marie-galantes* have become established in Ghana in West Africa (Hutchinson, 1949), but otherwise they are unknown beyond the New World. The *punctatums*, however, supplanted the old-established Asiatic cottons right across Africa south of the Sahara and are also to be found on the coasts of East Africa, on the islands of the Indian Ocean, and in South India.

In the two hundred years since they were first introduced (Hutchinson, 1949) they have undergone a remarkably extensive differentiation. In West Africa and nowhere else, true annual forms of *punctatum* have arisen, and in West African *punctatums* alone among the New World species of *Gossypium* is to be found high resistance and even virtual immunity to bacterial blight disease. The development of two new characters in a population in the course of two hundred years indicates a remarkably rapid differentiation. Neither has become universal, and the range of variation in the West African *punctatums*, in these as well as in other characters, is very wide.

A further line of development of some interest in *punctatum* is that which has given rise to the 'Hindi weed' of Egyptian cotton. *Punctatums* are common in the Sudan, whence the *barbadense* progenitors of Egyptian cotton were taken to Egypt. It was doubtless from the Sudan, therefore, that the first *punctatum* contamination of Egyptian cotton came, and among Sudan specimens of Hindi weed, all gradations are to be found between the perennial shrubby *punctatums* and the annual, rather larger-bolled Hindi as understood in Egypt. A study of this range of material indicates that the annual habit, rather larger boll, and rather better lint quality of the Egyptian Hindi have arisen by introgression from the Egyptian (*G. barbadense*) type, coupled with natural selection for success as a weed, that is for such advantageous characters as matching the host crop in duration and beating the host in seed germination.

Apart from the West African *punctatums*, all the annual forms of *G. hirsutum* belong to one race, to which the name *latifolium* is appropriate (Hutchinson, 1951). The original centre of the race was in central Mexico (the State of Chiapas), and from there it has spread over the territory of the other Central American races, and thence over the whole of the world's cotton-growing areas. The original stocks from Central America were predominantly photoperiodic, fruiting only in short days. Hence a large proportion of the *latifolium* material introduced into higher latitudes either failed to fruit altogether, or only produced small crops before the onset of winter. Forms capable of fruiting irrespective of day length were, however, selected in the southern United States, and have given rise to the section of the race to which the name 'Upland' cotton properly applies. The 'Uplands' have been distributed wherever cotton is grown, and besides providing the cottons of the United States cotton belt they now form the basis of all the commercial cotton crops of Africa outside the Nile Valley, all those of South America except Peru and north Brazil, all the Russian crop, and a considerable part of the crops of northern India and Pakistan. Race *latifolium* has made one other contribution to the world's cotton crops. Stocks of this race were taken by the Spaniards direct from Mexico to the Philippines, and thence to South-east Asia. Commercial cotton from Cambodia was so attractive to cotton spinners in Madras that seed was obtained and developed by the Agricultural Department in Madras under the name of Cambodia (Hutchinson, Silow & Stephens, 1947). Cambodia is now widely grown in South India, and has spread north and contributed much to the *latifolium* cottons grown as far north as Madhya Bharat. Having been taken to Asia direct from Mexico, the Cambodias were never subject to selection for ability to fruit in long days, and they still retain the short-day habit. Hence as they spread north in India they came into less and less suitable day-length conditions, and in the cotton areas of the Punjab and Sind they have been unable to compete with the day neutral true Uplands descended from introductions of 'New Orleans' and 'Upland Georgian' cottons made by the East India Company.

The Cambodias are characterised by a dense coating of long hairs on both leaves and stems. Hairiness gives resistance to the jassid pest, and the Cambodias have the best resistance known. Not only has this hairiness been a factor in their success in South India. There can be little doubt that the extreme hairiness of modern jassid-resistant strains of the African Uplands owes much to deliberate hybridisation or accidental crossing with Cambodia.

The Uplands have gained extensively by hybridisation with other races and species of the cultivated cottons. Modern high-quality Uplands owe their superiority to introgression from high-quality *barbadenses* (Kerr, 1951). High jassid resistance has been acquired from Cambodia, and high resistance to bacterial blight is being developed following hybridisation between West African *punctatum* and Nigerian Allen. It is likely that *Verticillium* wilt resistance in the United States was derived from *barbadense* also, and the proximity of American Egyptian crops to the Acala type Upland crops of New Mexico and Arizona no doubt provides the basis for yet further introgression into the Acalas of the western United States.

Reference must be made to the third allopolyploid species, *G. tomentosum*, which is wild and endemic in Hawaii. Such cytological work as has been done on it (Gerstel & Sarvella, 1956) affords no reason for regarding the species as having resulted from a separate occurrence of polyploidy. It is the least like a true cotton of all the allopolyploids, having seed hairs that are not differentiated into lint and fuzz, and that are only ⅜ inch in staple. The hairs are reddish brown, and are more nearly fully thickened than in the true cottons. Unfortunately the species is difficult to study. In the United States Cotton Belt it can only be grown in greenhouses where it is protected from winter cold. In the Sudan it dies out with leafcurl, and in Uganda it is destroyed by bacterial blight.

The three allopolyploid species of cotton cross freely, giving vigorous and fertile F_1 hybrids. In F_2 extensive genetic breakdown occurs, and a large proportion of the segregates are so unthrifty as to be virtually sterile, even with careful culture. With the spread of the two cultivated species, opportunities for crossing between them are frequent, as forms of *G. barbadense* and *G. hirsutum* are to be found growing in mixed crops under primitive agricultural conditions in many parts of the world. Highly vigorous, fertile plants intermediate between the parent species are not uncommon in such circumstances. They match artificially produced F_1 hybrids in all respects and there can be little doubt that they are natural F_1's. They have been recorded in the West Indies (Stephens, 1946) and in Ghana (Hutchinson, 1949) between *G. barbadense* and *G. hirsutum* race *marie-galante*, and are known to occur in the Sudan between *G. barbadense* and *G. hirsutum* race *punctatum*. They

37

are also of common occurrence in fields of Acala (*G. hirsutum* race *latifolium*) in the irrigated regions of the south-western United States where crops of Acala and American Egyptian cotton (*G. barbadense*) are grown in close proximity. Even *G. tomentosum* is not sufficiently isolated to escape hybridisation. Dr O. Degener (personal communication) forwarded in 1950 dried specimens and seed cotton from a single large plant found growing in a clump of *G. tomentosum* at Nanakuli on the island of Oahu, Hawaii. The plant was intermediate in appearance between *G. tomentosum* and *G. barbadense*, and from the quantity of specimens and seed cotton provided it must have been very vigorous and highly fertile. A progeny grown at Namulonge, Uganda, had all the characteristics of an F_2 of an inter-species cross. The plants varied enormously in size and vigour, many seedlings failed to develop into mature plants, and a great majority of the plants that reached maturity produced little or no fruit. There can be no doubt that the Nanakuli plant was a naturally produced F_1 between *G. tomentosum* and *G. barbadense*.

Stephens (1946), discussing the genetics and distribution of the 'corky' abnormality, has suggested that in the areas of overlap of *G. barbadense* and *G. hirsutum* in the West Indies, a sterility barrier is in process of development. Quite apart from this, the integrity of the species is maintained under conditions where crossing is possible, by the failure of the great majority of F_2 plants to contribute to later generations. Harland (1933) first set out the genetic basis of this situation. He showed that the genotype of a species must be regarded as built up as a balanced gene system. This internal balance is broken down when two species are crossed, and the disintegration following recombination between species genotypes in F_2 and later generations results in the failure of a large proportion of the segregates. The establishment in the offspring of inter-species hybrids of stocks combining the good qualities of the two parents is consequently a very slow, and often disappointing undertaking. In all successful material of inter-specific origin the basic genotype of one of the parent species has been re-established, the gene contribution of the other species being very limited.

An extensive and long-continued attempt was made at Barberton in South Africa to establish long-staple types with the productivity of Upland cottons by inbreeding the descendants of an Upland × Sea Island (*G. hirsutum* race *latifolium* × *G. barbadense*) cross. Successful commercial types were not in fact produced, largely because commercial types with the desired characters were bred from a cross between a long-staple Upland and the current South African Upland stock (an intra-*hirsutum* cross) before the products of the inter-species cross had been bred up to commercial standards. Nevertheless, it was evident in the later generations of the experiment that intermediate types had given place to forms that were in all respects within the range of variation of *G. hirsutum* race *latifolium*. Thus, selection for vigour, stability and fruitfulness re-established the genetic balance of one of the parent species.

There can be no doubt that introgression can and does take place successfully between species of the degree of differentiation existing in the commercial cottons, but all the evidence indicates that it is a long process. Barberton experience, and a survey of a long list of other inter-species hybrid projects that were abandoned, and of material reputed to have undergone intercrossing an unknown number of generations earlier, suggest that a period of twenty-five generations is about the length of time required to establish the products of introgression unless a system of backcrossing is adopted.

With backcrossing, gene transfer, which is the basis of introgression, can be achieved more quickly. Repeated backcrossing to one parent ensures the maintenance of the genetic balance of that species, and permits the integration of a portion of the gene content of the other species without the risk of a genetic breakdown. An extensive programme of gene transference on these lines was carried out by Knight (1954), who was interested in adding to the genotype of Egyptian types of

G. barbadense genes for resistance to bacterial blight from *G. hirsutum*.

Knight's experiments are of great importance for an assessment of the significance of introgression as a process of plant improvement. From observations on natural populations, it is often assumed that no more is required than some fertility in crosses between species growing together for introgression to go on. The analysis begun by Harland of the genetic nature of the species distinction in the cultivated cottons, and the precise study of the consequences of gene transference carried out by Knight in the Sudan, make it possible to define more exactly the true circumstances governing introgression.

In the first place, introgression does not normally occur unless there is some kind of a genetic barrier between the parent types. Between races of the same species in the cottons there is no genetic barrier and, apart from any selection exercised by man, where two races meet, the distinction between them vanishes as their genotypes dissolve in a common gene pool. This fusion can be seen in *G. barbadense*, for example, in the West Indies, in the forest regions of Nigeria, and in Egypt.

Where a species barrier exists there must, of course, be a sufficiently close relationship between the parents to allow of reasonably easy integration of genes from one species into the other. Again, cotton breeding supplies data not available elsewhere on the nature of the limits to introgression. Attempts have gone on for nearly twenty years to integrate into the New World allopolyploid cottons genes from the diploid *G. thurberi* that give high fibre strength. Kerr (1951), after twelve generations' work, maintained that he was 'convinced that in spite of the difficulties we shall ultimately be successful, but success will come only with time and a great amount of hard work'. Success appeared to be little nearer in 1954 after three more generations. Evidently the difficulties are indeed formidable, and the hope of integrating *G. thurberi* genes for fibre strength in a *G. hirsutum* genotype and obtaining a type competitive with the *G. hirsutum* parent must be judged slender. On the other

40

Spread of New World Cottons

hand, at least two genes—Knight's bacterial blight resistance gene B_{6m} and Harland's anthocyanin gene R_2—have been transferred from Asiatic diploid cottons, established by Knight in strains of the allopolyploid *G. barbadense*, and shown to be competitive with the parent *G. barbadense* stock. In neither of these cases would there be any possibility of introgression in nature, as the sterility barrier is too great, but they do show something of the genetic limits to introgression.

Introgression normally takes place where the parent species are differentiated to the degree found between cultivated species of *Gossypium* with the same chromosome number. Knight's single gene transferences of bacterial blight resistance from *G. hirsutum* to *G. barbadense* illustrate the mechanics of introgression, the conditions for success, and the limitations to which it is subject. In breeding stocks it is easy to preserve material of good genetic potential but poor phenotype. In nature such material would be eliminated. Hence an important factor in the preservation of the material on which successful introgression in nature is based is heterosis. The inter-species F_1 and a proportion of its backcross derivatives will be among the most vigorous and productive plants in the population. In later backcross generations, with the re-establishment of the genotypic balance of the recurrent parent, types will arise carrying some genes from the donor parent which will be competitive without the advantages of heterosis.

Knight's work demonstrates the importance of a real advantage from the introduced genes if introgression types are to succeed. Under the rigorous blackarm control exercised in the Sudan Gezira, genes for bacterial blight resistance conferred an advantage only in exceptional epidemic years, and though the value of the resistance was undisputed as an insurance, and as a safeguard if it became necessary to allow a less strict system of plant sanitation, it was in fact difficult for Knight to demonstrate under those conditions a selective advantage for his resistant stocks. This is a point to be borne in mind when considering the possibility of introgression in natural populations where precise comparisons of selective advantage are hard to make. Introgression involves the progressive synthesis of a new type, and it will only go on if in each successive generation the developing type can maintain itself in competition with the parent stock.

The value of carrying out work such as this in a crop plant should be noted. It is not difficult to transfer gene material across fairly wide species barriers, and to synthesise new stocks morphologically within one species but carrying some genes from another. The synthetics may well be normal in appearance, fully fertile, and easily maintained in the experimenter's cultures, but it is only when such stocks have to compete in the full rigour of field trials that an objective test of survival value is obtained.

4

THE PATTERN OF EVOLUTION IN THE COTTONS

A striking feature of the evolution of the cottons is the association between variability and population size. This is well demonstrated by the great contrast between the cultivated cottons and the wild species of *Gossypium*, particularly those of the *D* group in the New World. The former are wide ranging, with enormous populations and very extensive variability. The latter are generally confined to small areas, with small populations, and with very little apparent variability. Even the demonstration of appreciable variability in comprehensive collections of *G. thurberi*, the widest ranging wild *Gossypium* in the New World, does not materially affect the contrast. In comparison with the variability to be seen among the crop plants, that to be found in *G. thurberi* is infinitesimal.

Evolution in the Cottons

Numerous examples of the relation between population size and diversity are to be found within the cultivated cottons also. The geographically and agriculturally restricted races *morrilli* and *palmeri* in *G. hirsutum* are relatively uniform, whereas the wide-ranging and agriculturally successful races *marie-galante*, *punctatum* and *latifolium* are extremely diverse. Even within a race, the scattered house-yard population of *punctatum* in the Bahamas is far less variable than the widespread *punctatum* population of the house-yards and small cultivations of the southern margin of the Sahara.

This diversity among the cultivated cottons must have arisen since they were domesticated. At the time when the ancestor of *G. herbaceum* race *africanum* was adopted as a crop plant, the variability available must have been very small indeed. Crossing with other stocks can be excluded as a source of the variability, as no other species or race was then available that could conceivably have contributed anything to the cottons. Thus the variability must have been of internal origin.

In the final analysis, new variability can only arise by mutation, and the problem is therefore to account for the occurrence and survival of mutations at such a rate as to explain the enormous diversification of the cultivated cottons within the very short period (in evolutionary terms) since their domestication by man. Mutation rates are very difficult to estimate. Nevertheless, Fisher (1930), who discussed such information as was available, showed that the overall rate of mutation is adequate to account for the development of diversity. The problem therefore becomes that of determining the circumstances under which diversity arising by mutation is accumulated. Fisher (1930) confirmed the conclusion reached by Darwin that variability within a species is greater in species with large populations than in those with small populations. He showed that

in an increasing population mutations possessing no selective advantage, or indeed mutations at a selective disadvantage, provided this is less than the rate of increase of the species as a whole, will have a finite chance of avoiding extinction; while with a declining population, even mutations possessing a slight advantage, if this is less than the rate of decrease of the species, will be in a worse position than neutral mutations in a species of stationary size. In consequence growing populations receive greater accessions to their variability than stationary populations, while declining populations receive less; and if the intensity of selective action is the same in both cases, we may expect growing populations to grow more variable, and declining populations to become less so by a process which is distinct from the effect of population size itself upon variability.

Thus there is a mechanism whereby an increase in variability might follow from domestication. Fisher continued, 'The scope of this cause is limited by the actual rates of increase or decrease of natural populations, and I suppose that such changes are seldom so great as an increase of one-hundredfold in 10,000 generations, or about 1 in 2000 in each generation over such a period.' From this he concludes that the advantage or disadvantage of mutations the surviva of which is affected by those changes in population size must be of the order of 1/2000 or less.

Circumstances in a crop plant are often of a very different order. An estimate of the rate of increase that takes place when a new crop plant is introduced may be obtained from the history of the cotton crop in Uganda (Tothill, 1940). Acreage figures and spacing are very uncertain, so any estimate of numbers would be subject to considerable error. Yield per acre may be taken as fairly constant over a period, and an estimate of the rate of increase of the population may be obtained from the annual crop production figures, which are based on actual sales, and have therefore a precision lacking in acreage estimates. In the thirty years from 1906 to 1936 the crop rose from 500 bales to 316,000 bales, giving an estimated increase in crop population of about 600-fold in thirty years. This corresponds to a rate of increase of nearly 25 per cent per annum. A rate of increase of this order or even greater might well occur in a newly domesticated crop, once its desirability was accepted. Thus, the conditions for the establishment of newly occurring variability would be far more favourable in the circumstances attending the domestication of a new crop plant than in natural populations in which the rate of increase might never exceed that postulated by Fisher.

Fisher considered changes in population size on the assumption that 'the intensity of selective actions is the same'. It may be argued, however, that changes in population size are indissolubly linked with changes in selection pressure. A population increases in size when the selective sieve lets through more individuals than got through in the previous generation. Hence genotypes that would fail in a stationary population have a chance of survival in an expanding one. Moreover, the precise estimates of selective advantage appropriate to the calculation of the survival and spread of genes in a static population are hardly relevant to a discussion of the development of variability in an expanding species. In the field, a great range of environmental conditions is to be found, and it is to be expected that there will be considerable variation in the survival value of a gene according to the environment in which it occurs. Hence the survival of a new mutant gene will depend, not on a small balance of advantage to the species as a whole, but on its fitness in the particular environmental niche in which it occurs, and in an expanding species, the greater the diversity of the environments into which it spreads, the greater will be the amount of newly arising variability that will be conserved.

The diversity of environmental circumstances is likely to be greater in crop plants than in wild populations, since man extends the range of his crop plants beyond their natural limits by destroying vegetation that would normally compete with them. Thus in Antigua, for instance, the secondarily wild *G. hirsutum* race *marie-galante* is confined to one habitat, the seasonally wet valley bottoms, where it occurs as scattered shrubs in a scrub vegetation, whereas the Sea Island (*G. barbadense*) cultivated in the same island is to be seen in large crop populations in habitats experiencing a considerable range of rainfall and on a wide diversity of soil types.

Genetics and Cotton Improvement

The way in which diversity in habitat may favour the survival of mutant genes that are not generally favourable is well illustrated by the naked lintless mutant of Upland cotton. This occurs not infrequently in American Upland cotton, in the sense that perhaps a dozen or more naked lines have been picked up and established as genetic stocks in culture. They are, of course, immediately eliminated from the commercial crop. A similar, very likely the same, lintless type was found by Martorell in 1943, growing in a small population of *G. hirsutum* race *punctatum* in Puerto Rico (Hutchinson, 1944). In these secondarily wild *punctatums*, the lint is not used and it may well be no advantage, so there is no reason why the mutant should not persist. That it probably has been long established is indicated by the fact that Todaro figured it (as *G. racemosum* Poir.) in Puerto Rican material more than eighty years earlier (Hutchinson, 1944).

Evidently the circumstances of the development and spread of a crop plant are such as to favour the development of variability, and the escape of a crop plant from the uniformity characteristic of a wild plant oppressed by the selection pressure of a limited environmental niche is entirely explicable.

A distinction must be drawn between the crop populations of unspecialised agriculture and those of modern farming systems in which planting seed is provided by seedsmen or plant breeding institutions. Population growth in the former provided favourable circumstances for the establishment of new variability, and the breeding system was such as to ensure its maintenance. In the latter, population growth is enormous in each generation, but the organisation of the seed supply is such that the main crop is regularly replaced by the products of a breeding system based on a very small effective population. The seed replacement scheme for V 46, the strain of Sea Island Superfine V 135 now grown in St Vincent, may be taken as an example. Each year twenty plants are selected and grown in progeny rows. Their produce is bulked on the Experiment Station in the following two years, and grown on an estate in

46

the fourth year. In the fifth year sufficient seed is available to plant the whole island, and in the sixth year, that stock is discarded and crushed for oil, and the next wave follows on. There may be 3000 acres at 7500 plants per acre in the island crop, so the twenty original plants multiply up to 20 million in five years, a phenomenal rate of increase, but there the matter ends, and the effective breeding population is twenty plants. It is not surprising that this system of breeding and multiplication has been highly successful in maintaining the extreme uniformity considered necessary for the production of commercially acceptable superfine Sea Island cotton. On the other hand, it is not to be expected that the stock will be susceptible of improvement. It is, perhaps, of interest to note that when in the past this, or any of the other similar West Indian Sea Island breeding systems, was relaxed, and the rapidly multiplied population allowed to breed for several more generations, complaints of 'degeneration' soon arose. Whatever else 'degeneration' may mean, it certainly involves an increase in diversity.

Turning from the origin of diversity to its exploitation in the emergence of new agricultural forms, the history of the distribution and differentiation of *G. hirsutum* may be considered. Studies of variability in crop plants by Vavilov and others have shown that in the later stages of their development there is often great diversity at the centre of the species, and a considerable degree of uniformity around the periphery of its geographical distribution. It was in the belief that the greatest variability was to be found at the centre of origin of the species that the collecting expeditions of Richmond, Stephens, and Ware and Manning to the centre of origin of *G. hirsutum* (Hutchinson, 1951) were carried out. They revealed indeed great diversity in Central America, but it was very largely diversity between perennial races rather than diversity within races, and in any case involved characters that are not relevant to the breeding of modern annual commercial cottons. It now appears that the diversity in Central America arose from only the first of a

Genetics and Cotton Improvement

series of stages in the development of *G. hirsutum*. This stage was to a large extent stabilised by the partition of the diversity between the seven Central American races. The second stage was marked by the extension and differentiation of the two perennial races *punctatum* and *marie-galante*, when *punctatum* spread through the coasts and islands of the Gulf of Mexico and later across West Africa from Senegal to the Red Sea, and *marie-galante* through the Spanish Main and the Caribbean islands. The third period of development followed the introduction of the annual race *latifolium* into the southern United States, the establishment of the American Upland cotton crop, and the spread of the Uplands into Asia and Africa. Each of the three races that spread beyond Central America acquired a fresh range of variation in its new areas of distribution. In *marie-galante* the biggest and most tree-like forms in the genus are to be found on the Spanish Main and the Caribbean islands. In *punctatum* the annual habit has been developed and a high level of resistance to bacterial blight built up in the two centuries or so that the race has been in Africa. In *latifolium*, a level of productivity beyond anything previously known has been achieved in the United States, and an enormous range of types developed that are adapted to agricultural and climatic circumstances as diverse as those of the irrigated regions of Arizona, California, Pakistan and the northern Sudan, the warm temperate and sub-tropical rain-fed areas of the United States Cotton Belt, São Paulo in Brazil, many parts of India, and the Transvaal and Moçambique, and the rain-fed tropical regions of Central America, and Uganda, Nigeria, and the Belgian Congo. Nor is this all. Diversity in these three races has been further augmented both by gene exchange between them, particularly in recent years in Africa, and by introgression from *G. barbadense*, which has resulted in important accessions of quality genes to both *marie-galante* and *latifolium*.

Thus it is evident that the generally accepted picture of great diversity at the centre with progressively declining variability towards the periphery does not hold. It is convenient to illustrate the situation by reference to *G. hirsutum* because the evidence in this species is particularly clear. The situation is similar, however, both in *G. barbadense* and in the two diploid Asiatic species, *G. arboreum* and *G. herbaceum*. It is therefore necessary to consider in what respect the breeding system of the cottons differs from that of a crop plant such as wheat, in which the evidence for the decay of variability from species centre to periphery is well established.

It has been pointed out (Hutchinson, 1951) that the crop plants in which the Vavilov effect has been demonstrated are self-fertilised, at least over the greater part of their peripheral range, whereas all the cultivated species of *Gossypium* enjoy a considerable degree of cross-pollination over all, or nearly all, of their distribution. It has been shown by Rick (1950) that the tomato, which is self-fertilised in its European and North American cultures, is largely cross-pollinated in its original area in South America. Moreover, Howard & Howard (1909) showed that in the Punjab in northern India and Pakistan wheat is frequently cross-pollinated. Now if a cross-pollinated crop plant spreads into an area in which there are no suitable pollen vectors (as with the tomato), or if for any other reason self-pollination becomes advantageous, as, for instance, in ensuring complete pollination and the full set of seed on which a high agricultural yield depends, a change in the breeding system will follow, and a population in which there is free gene exchange will be replaced by one in which the variability is partitioned between a number of homozygous lines. Thus the variability will be made up of what was inherited from the immigrant population plus what may become established by new mutation, but there will be no opportunity for assortment of the newly arising variability into new and possibly favourable gene combinations. In such circumstances the Vavilov effect may be expected to develop, and recourse to the original areas of the species, where gene exchange continues, may well offer the best prospect of finding new and favourable gene combinations.

On the other hand, where gene exchange goes on throughout the species, there is every opportunity for the occurrence of recombinations of newly arisen mutants, and new variability may be expected to contribute to the synthesis of new characters that will increase the adaptation of the race to the new conditions in which it is grown. Moreover, incipient specialisation will be balanced by gene exchange as the spread of the crop brings the new forms into contact with each other. The way in which this pattern of evolutionary development has proceeded in the peripheral areas of the distribution of *G. hirsutum* may be illustrated by accounts of the development of the annual habit, and of genetic resistance to bacterial blight, the jassid pest and *Verticillium* wilt.

Gossypium is essentially a genus of perennial shrubs. Flowering takes place in the latter part of the rainy season each year and the fruit is matured in dry weather. In the primitive perennial forms of the cultivated species the onset of fruiting is governed by three major factors. In the young plant fruiting branches are not formed at all until a certain number of nodes have been laid down on the main stem, and this node number is a genetic characteristic of the stock. Thereafter, fruiting branches will only develop if the day length is below a critical level, which is probably rather less than 12 hours. Finally, even if fruiting branches are formed and young flower buds develop, these buds are shed at an early age if copious supplies of water are available to the plant. Thus a perennial cotton only fruits if it is big enough, if its photoperiod is right, and at the season of the year when the rains abate and the dry season sets in. The first step in the production of an annual cotton was the selection of types in which the first fruiting branch was formed low on the main stem. Such types produced a crop in the first year of growth, provided they remained in the same region as their forbears, where their photoperiodicity and their water response were in balance with the local seasons. From such a type sprang on the one hand the *punctatums* and on the other the *latifoliums*.

Evolution in the Cottons

In the *punctatums*, spreading north through Yucatan and Cuba to the Bahamas and Florida, the short-day habit was out of balance with the local climate of the more northerly latitudes, and disappeared under natural selection. The shrubby perennial habit was retained, but the low node number resulted in the production of a considerable crop in the first year. From the Bahamas *punctatum* type, early fruiting but shrubby, almost or entirely day neutral, but with a strong tendency to shed flower buds in wet weather, it is an easy step to the perennial *punctatums* of West Africa. But in West Africa a further step has been taken. Even lower node numbers have been selected, and the strong vegetative growth at the lower nodes characteristic of the shrubby *punctatums* has been eliminated, and early, prolific, annual forms have arisen that produce a crop and are virtually finished in six months.

In the *latifoliums*, transfer from Mexico to the southern United States was only successful with a small proportion of types in which the photoperiodic response was sufficiently weak to allow the development of a crop before the onset of cold weather. The selection pressure both for reduced vegetative growth and for the elimination of the photoperiodic response was very strong. Moreover, any tendency to shed flowers in wet weather was deleterious, and in consequence the Upland race rapidly developed the full annual form, in which vegetative growth is strictly limited, fruiting branches are formed very early, and fruit production goes on regardless of day length or of the duration of the wet season.

These are two of the lines of development that have led to the emergence of annuals in a genus of perennial shrubs. They are two among many. The annual habit has arisen in *G. barbadense* in the Sea Islands and Egyptians, and in this case the development is known to have started no longer ago than 1785, with the introduction of *G. barbadense* into the Carolinas and the beginnings of the Sea Island crop. Among the Old World cottons, the annual habit was confined to the extreme northern periphery of the distribution of *G. herbaceum* when Marco

Polo went on his travels. Since that time it has been developed independently in at least three lines in *G. arboreum*, and annuals of both species have spread until they have virtually excluded the perennials from commercial crops. An interesting illustration of the genetic independence of two syntheses of the same character was provided by a study of the genetics of the annual habit in crosses between two annual types, *G. arboreum* race *bengalense* and *G. herbaceum* race *persicum*. Within each species there is a close association between habit and node number, those perennial forms in which the first fruiting branch arises high on the main stem having numerous vegetative branches and a bushy habit, while the early annual forms combine early appearance of the first fruiting branch with few or no vegetative branches, giving a single-stemmed almost herbaceous habit. An inter-species cross between two early fruiting, single stemmed forms gave in F_2 a much increased range in node number, and a wide range in the development of vegetative branches, and moreover the correlation between number of vegetative branches and node number was much reduced. Thus the pattern of evolution of the annual habit had been the same in the two species, but the genetic threads with which it was woven were different.

Bacterial blight (*Xanthomonas malvacearum*) of cotton is a disease of Old World origin (Knight & Hutchinson, 1950), and its impact on the New World cottons dates from post-Columbian times. Thus any gene differences affecting resistance to the disease may be assumed to have been without selective advantage in the New World until 400 years ago. There are now to be found, however, *punctatums* of the West African acclimatised group that are virtually immune to the disease. The genetic basis of resistance, and the gene combinations that raise the level of resistance to virtual immunity, have been extensively studied by Knight (for references see Knight, 1954), and the history of the development of resistant types in the New World cottons has been set out by Knight & Hutchinson (1950).

Evolution in the Cottons

Knight worked at Shambat in the northern Sudan where the relation between the disease organism, the host and the environment was such that he was able to get an extremely clearcut differentiation between hosts differing in resistance. He isolated six genes with sufficiently large effects to be individually identifiable under his conditions, and showed that there were also genes of small effect that could not be separately studied. He demonstrated that these genes were in general cumulative in action, and plants having a high degree of resistance carried not less than two of the larger genes, and often a number of those too small to be separately studied.

The importance of these bacterial blight resistance studies is twofold. First, Knight's genetic analysis is the most complete so far made of the make-up of a multifactorial character, and secondly, Knight and Hutchinson's survey has shown how such a character is built up under selection. Knight divided the genes concerned into major genes and minor genes, but such a division can only be arbitrary. Major genes were those for which he could follow the segregation individually. Minor genes were those that could not be so followed. His major genes differed considerably in the magnitude of their effects, and his B_1 for instance had so small an effect that it could only be followed under favourable circumstances, even at Shambat. Workers on other stations in Africa have consistently failed to achieve a sufficiently uniform incidence of the disease to enable them to follow the segregation of any of Knight's major genes. Hence his division into major and minor genes depended more on the uniformity of the environment in which he worked than on the magnitude of the gene effects, and it may be concluded that resistance to bacterial blight is governed by a number of genes that vary considerably in the magnitude of their individual effects and that are in general cumulative in their action. Knight's very great achievement is that he seized the opportunity presented by an exceptionally suitable environment to conduct an extremely comprehensive analysis of a multifactorial character.

Genetics and Cotton Improvement

From Knight and Hutchinson's survey it appeared that resistance to bacterial blight is rare among cottons on the American continent. Only among *punctatums* from the Bahamas were plants with appreciable resistance discovered. That genes giving resistance do arise from time to time in American material is indicated by the discovery of the resistance gene in Stoneville 20 (Weindling, 1948), and by the success of very extensive field surveys in revealing resistant plants in American commercial varieties (Brinkerhoff, Green, Hunter & Fink, 1952). Without the selective advantage enjoyed under conditions of heavy blight attack, they do not appear to have spread through the population, and it may even be that in the absence of blight they are at a disadvantage.

The discovery of some resistance in *punctatum* in the New World indicates the starting-point for the development of resistance among the *punctatums* introduced into West Africa. The extent of the development was, however, far beyond anything that could have been anticipated from the modest degree of resistance found in a proportion of the *punctatum* stocks in the Bahamas. In less than 200 years, virtual immunity has become common in West African *punctatums*. It seems likely that this has been possible because of the occurrence in West Africa of a mutation at the locus termed B_3 by Knight. Knight & Hutchinson (1950) showed that where Upland cottons in the Old World have acquired some degree of blight resistance they have had in the past direct contact with *punctatum*. Such Upland resistance is always based on Knight's B_2, with or without B_1. Only in West Africa is high resistance based on B_2B_3 to be found. Thus it may be supposed that the early *punctatum* introductions carried B_2, and were the donors of B_2 to Uplands in various places. Then in West Africa B_3 arose in the stock in which B_2 was already present in some individuals. The B_2B_3 combination would soon arise by recombination, and among plants carrying B_2B_3, selection for minor resistance genes, either already present or newly arising by mutation, would give rise to the virtually immune genotype now to be found with considerable frequency in the West African savannah region.

In contrast to the development of bacterial blight resistance in West African *punctatums* under natural selection, the establishment of resistance to the jassid pest (*Empoasca* spp.) in African Uplands has been the result of a well-planned and long-continued programme of plant breeding. The association between leaf hairiness and jassid resistance was known when the Barberton station of the Empire Cotton Growing Corporation was established (Parnell, 1925). Selection for resistance was based on response to jassid attack, but as the years went by and the understanding of the relation between hairiness and resistance grew, more attention was paid to the direct estimation of hairiness and less to the impact of the insect pest. Finally, Parnell, King & Ruston (1949) showed that length and density of leaf hair were in fact the direct cause of resistance to jassid.

The *latifolium* cottons of Central America include a considerable proportion of plants with some leaf and stem hair, but in general the hairs are neither long enough nor dense enough to give good jassid resistance. The modern Upland cottons of the American Cotton Belt have extremely sparse leaf hair. There is, however, good reason to believe that the varieties of the first decade of the present century that provided the basic stocks for Africa contained at least some moderately hairy plants. Knight (1952) has recently shown that all such types carry one of two genes for hairiness, and that intensification either in length or in density depends on the selection of genes modifying one of these basic genes. In West African Upland material, selection of this type gives stocks with good hair density, but only moderate length, and hence only a fair degree of resistance. In South Africa and East Africa, on the other hand, a degree of resistance that approaches immunity has been achieved by the selection of long as well as dense hair. It appears probable that the hair-length factors available in South and East Africa that are not available in West Africa have come from Cambodia. The Cambodias, both the commercial stocks from South India and cottons from small cultivations in the Philippines, are characterised by a dense coat of long hairs, and are virtually immune to jassid. Jassid is a serious pest in India, and may well be in Cambodia and the Philippines. Cambodia was grown at Barberton in the 1924–5 season and for some years following, and was also grown on experiment stations in Tanganyika and Uganda. Hence the opportunity arose for introgression into the African Uplands and then of the development of a degree of resistance not yet attainable in West African Uplands derived from the same United States stocks.

The way in which these characters have been synthesised suggests a mode of attack on a similar problem that is not yet completely solved. The occurrence and spread of *Verticillium* wilt on cotton in various parts of the world has been met by selection for resistance, but in Upland cottons no fully adequate resistance has yet been built up. Partial resistance has been developed by selection independently in at least three Upland stocks, the Acalas of the south-western United States, some of the older and now obsolete varieties of the south-eastern United States, and in the B 181 stock in Uganda. There is good reason to hope that at least some of the genes involved will be different in the different stocks, and it may be possible to synthesise from the three a degree of resistance higher than anything yet achieved.

The sequence of events in the development of a new character may now be set out. The raw material is the unorganised variability in the crop population. Under selection, new and superior genotypes are favoured as they arise, and the process of response to the new selective forces that give the new character its advantage continues so long as there remains variability in which selection can operate. When the limit is reached, further progress is only possible following the occurrence of suitable new mutations, or the transfer of suitable gene material by introgression from other stocks. In either case, the possibility of gene exchange through inter-crossing is essential for progress. It is an interesting commentary on current plant-breeding techniques that in the desire to achieve uniformity the plant breeder favours close inbreeding, which is calculated to limit most seriously the prospects of obtaining new and superior genotypes.

5

BREEDING SYSTEMS

The prospect of improvement in breeding work lies in the selection and propagation of superior material out of a heterogeneous population. There can be no breeding progress without selection, and selection becomes ineffective when variability is exhausted, or falls below the level at which it can be effectively detected. There has been much discussion of the merits of diverse systems of breeding, but little attention has been paid to the relation between the nature and extent of the variability in the material under selection, and the techniques by which it may be most effectively exploited. This relationship is of such importance that it is appropriate first to consider the kinds of variability with which the breeder may be faced.

The extent of the variability in stocks available to a plant breeder depends on the previous history of the crop and on the geographical range sampled. In an unselected country stock, or 'land-race', such as is commonly encountered in under-developed territories, the crops that can be sampled within convenient reach of an experiment station will often provide enough material for some years' work. Other biotypes of the species, and the wide range of diversity to be found in hybrids between such biotypes, may then be sampled, and this range of variation, within the limits of the species and therefore in a single gene pool, is usually sufficient for the needs of all but very long-term breeding projects.

There are occasions when all the characters the breeder wishes to assemble are not available in one species, and the problem then arises of combining gene material from distinct gene pools. The existence among the cottons of distinct species that may be crossed freely, even though they give genetically unbalanced progeny in later generations, has made possible extensive studies of the prospects of combining material from different gene pools. Harland (1939) first demonstrated the genetic consequences of segregation in crosses between inter-fertile species. He concluded that the establishment from an inter-species cross of vigorous, stable races, capable of main-taining themselves in cultivation, depends on the selection of material genetically very closely allied to one or other of the parent species. He predicted that it would be possible to include in a new stock resembling one parent, a small portion of the gene content of the other, but attempts to establish types intermediate between the parents would result in weak and unproductive plants.

Much of the genetic research conducted by the staff of the Empire Cotton Growing Corporation stemmed from this fundamental concept. It was obviously essential for intelligent cotton breeding to determine the number, nature, and limits of the gene pools to which the cultivated cottons belonged, and the rationalisation of cotton breeding began with the delimita-tion of the four species of cultivated cottons. Subsequent work on the distribution of variability within the four species (Silow, 1944; Hutchinson, 1951) gave precision to breeding projects, surveys of the distribution of variability in respect of particular characters (Knight & Hutchinson, 1950; Hutchinson, Knight & Pearson, 1950) led to the identification of stocks of direct plant-breeding significance (Wickens in Namulonge P.R., 1951), and the successful transfer of genes for blackarm resistance from *G. hirsutum* to *G. barbadense* (Knight, 1954) provided a practical demonstration of the feasibility of adding a small part of the gene complement of one species to the genotype of another without damaging its genetic balance.

The genetic analysis of the cultivated cottons led naturally to the application of genetic principles to cotton breeding. This was undertaken at Indore in Central India in 1933. At that time plant breeding in India was dominated by Johannsen's 'pure line' concept, and in all the early crop improvement work the first stage was to sort as wide a range of material as could be assembled into true-breeding 'types'. Segregating material was regarded as exceptional and undesirable, and was either 'purified' or discarded. Then the pure-breeding types were described and regarded as representing the range of material available in the crop population. These types provided the stocks on which the first improvement in crop production was based. Among them were high-yielding lines and disease-resistant stocks that laid the foundation of crop improvement in India. The pure-line theory so dominated thought at that time, however, that the importance of gene assortment as a source of new material of agricultural value, and the possibility of maintaining heterozygous stocks in which such assortment could proceed, were overlooked.

At Indore, the use in progeny row breeding of modern experimental techniques involving replication and randomisation was demonstrated (Hutchinson & Panse, 1937), and the way was opened to the measurement of the genetic variance on which improvement by selection depends. The direct relation between genetic variance and the prospect of advance in plant breeding (Hutchinson, 1940), and the loss to the breeder from 'the misguided selection of lines with low variance' (Hutchinson, Gadkari & Ansari, 1938) became accepted as fundamental concepts in a new approach to breeding problems.

The genetic principles involved, and the statistical tools with which to make use of them in practical breeding, were later developed by Manning, first with Sea Island cotton in the West Indies (Hutchinson & Manning, 1943), and more recently in Uganda with BP 52 (Manning, 1956*b*). The principles of a statistically controlled breeding system were set out by Hutchinson & Manning (1943) thus: 'In a well-designed breeding project [the progeny row system] can be made to yield information on the selection exercised, the improvement gained, and the possibility of further response.'

Studies of the efficiency of selection in progeny row breeding (Hutchinson & Kubersingh, 1936; Hutchinson & Manning, 1943) showed that the selection pressures actually exercised in a breeding project are often quite low. Partly for this reason, and partly because of the fundamental importance of assessing the comparative values for crop improvement of a number of plant characters, Manning (1956b) took the discriminant function technique (Smith, 1936), and developed from it a selection index. Having calculated the selection index for each progeny, a selection differential is set, and only those progenies with an index that passes the limit set contribute to the next generation. Under this system the selection pressure to be applied is determined precisely, and can be maintained from one generation to the next. Moreover, once a decision has been taken on the characters to be included in the selection index, the weighting of those characters and the selection of material for propagation is entirely objective. A prediction can be made of the advance to be expected, and the prediction checked in subsequent generations, and thus a check is kept on the extent of the residual variance on which the prospect of further advance depends.

The choice of a selection technique for any particular breeding project should be governed by an assessment of the variability available in the parent stock. With very diverse material, a large initial advance may be expected from a rapid general survey of a large amount of material, a technique which Mason (1938) called 'primary selection'. In the more uniform products of such a primary selection project, progress is more likely to be achieved by careful assessment of a smaller range of material under controlled breeding, that is, by pedigree selection.

Breeding Systems

Mason first distinguished primary selection as field selection in a large commercial crop, as distinct from the more usual method of selecting in small plots under controlled breeding on experiment stations. He contended—with some justification—that the more striking advances in plant breeding had usually resulted from primary selection, and not from orthodox breeding techniques, which he termed secondary selection. Examples of highly successful primary selections are very common in the history of cotton breeding. All the early advances in the breeding of Egyptian cottons resulted from primary selections. Farther south in Africa also, U 4 and BP 52 originated in this way, but both of these were later improved by pedigree selection.

These two varieties provide good illustrations of varietal improvement by selection techniques of steadily increasing precision. U 4 was bred at Barberton in South Africa by Parnell and his associates (Parnell, 1935; Macdonald, Fielding & Ruston, 1940). The primary selection, the original U 4, selected out of a very variable crop population of Upland cotton, was so great an advance on current stocks as to replace entirely the material from which it sprang, and to spread north into Southern Rhodesia, Nyasaland and the Belgian Congo, and east into Portuguese East Africa. The variability within it was great, though not of the same order as that in the original material. It was further improved by pedigree selection in South Africa and Southern Rhodesia, and an interesting sequence of strains was issued over a period of years, improved performance being accompanied by reduced variability and in general by increasing specialisation to the local environment. Thus, while the early Barberton U 4 strains rapidly replaced all other material in Southern Rhodesia, later Barberton stocks could not compete in Rhodesia with locally selected material of the same U 4 origin.

When the greater part of the available diversity had been exploited, further progress was sought at Barberton in the development of greater jassid resistance by hybridisation with other stocks of Upland cotton. The sequence of primary selection followed by pedigree selection was repeated to exploit the newly created variability, and these inter-strain hybrids within *G. hirsutum* race *latifolium* yielded stocks that combine very high jassid resistance and superior quality with a level of yield that is agriculturally acceptable in the areas for which they were bred.

In Uganda, a primary selection was made by A. L. Stephens in 1928. Nye carried on pedigree selection in it, and in four generations developed a superior derivative known as BP 52. BP 52 was first multiplied for commercial issue in 1938 (Jameson & Thomas, 1952). In BP 52 the prospects of advance lay chiefly in the selection of better yielding strains with improved minor quality characteristics while maintaining the same broad quality class. For this objective there was no stock with which it was likely to be advantageous to make a cross, and when Manning took over the BP 52 stock in 1945, he set out to exploit the variability still available within it by statistically precise pedigree selection methods. The selection index technique (Manning, 1956*b*) is designed to determine the heritability of each of the characters considered to contribute to the yield of a single plant progeny, and so to combine the estimates of heritability as to provide a single estimate of 'net worth' on which progenies can be selected. Heritability is the ratio of the genetic to the total variance, and estimates of genetic variance and total variance are obtained from replicated progeny row trials.

It should be noted that the heritability for any given character depends on the nature and extent of the segregation for that character in that stock in that generation. Hence it is necessary to determine heritability from the actual material under selection in each generation. The selection index so computed gives the most effective weight to each of the characters considered for that generation, and heritabilities determined in one generation are not necessarily a guide to selection in a later generation of the same material, let alone in another stock.

Breeding Systems

Manning (1956b) has shown, in fact, that calculation of the selection index has resulted in weighting the components of yield differently in successive generations of the BP 52 stock, partly because the exploitation of the variability in one generation affects the variability in the next, and partly because the environmental influences on the component characters differ in different seasons. Thus it may happen in some years that environmental fluctuation in number of bolls per plant is so great that the genetic variation is completely masked, and the heritability for that character is nil. In other seasons, however, environmental fluctuation may be less, and genetic diversity then shows up, giving a heritability large enough to justify the exercise of selection. The success of the precise statistical assessment of heritability has been such that by the selection index method a yield advance of the order of 4 per cent per generation has been achieved over eight generations, and there is as yet no evidence of a decline in the rate of advance.

The selection index technique may be regarded as the limit of refinement of selection methods. Its use makes possible the detection of any genetic variability in excess of the environmental variability, and its precision is therefore determined by the extent to which environmental variation can be controlled. This in turn depends on the multiplication rate of the organism under study, since the precision of progeny trials is limited by the number of progeny per parent available for test.

The development of precision in breeding techniques is an advance of the greatest importance. In any breeding project there must come a time when the material for further advance is exhausted, and a technique by which a constant check is maintained on the rate of improvement makes it possible at any time to assess the value of the return obtained in relation to the effort expended. When no further worthwhile improvement is to be expected, the stock can be put on to a maintenance basis, and materials for further advance sought in related, more variable stocks, or in the products of hybridisation.

Breeding may be regarded as an exercise in the management

of variability. Only while variability remains is there a prospect of genetic advance, but the effect of selection is to reduce variability. Thus the breeder's aim should be to shift the population values in the direction he desires, while maintaining so far as possible the variability which is the source of further improvement.

In all crop plants in which continuous self-fertilisation either takes place in nature or can be imposed in experimental culture without serious loss of vigour, it has long been customary to release for general use strains that are as close as possible to complete homozygosis. In modern, large-scale agriculture such a policy has the advantages of ensuring uniformity in growth and in the quality of the product, ease of identification, and ease of detection of contamination. In plant-breeding practice, however, it leads to an unreasonable emphasis on 'purity', and even to the precipitate elimination of the greater part of the variability in which lay the potentialities of future improvement.

Though the reduction of variability is no longer a primary objective in breeding, nevertheless line breeding, generally with self-fertilisation, is so universal a technique that it remains an incidental consequence of almost all plant-breeding projects. Moreover, since inbreeding restricts reassortment, breeding systems involving self-fertilisation do not fully exploit the potentialities of the gene pool.

Means of conserving the diversity of a breeding stock have received little consideration. Manning (1956b) has shown that the decay of the genetic variance, even with rigorous self-fertilisation, is surprisingly slow in both Sea Island and Upland cotton stocks. This is no doubt partly due to unconscious selection of heterozygous progenies for propagation, either on account of heterosis or because such progenies offer a better choice of material. With precise statistical techniques such as those employed by Manning, it has become possible to pick out those progenies in which segregation is still going on, and to select deliberately in heterozygous material. Little is known of the possibilities of selection in a gene pool composed of randomly interbreeding individuals, but it may be suggested that the rate of change under selection in a panmixis population of a few hundred individuals is well worth studying. Such panmixis populations are easily set up with cotton, where extensive cross-pollination can be ensured by transferring pollen from flower to flower with a camel-hair brush.

In planning any such departures from orthodox progeny row breeding techniques, there is much to be learnt from the data and the theories of population genetics. The rate of change of gene ratios under selection, the effect of random chance on gene survival in small populations, and the effects of population structure on rate of change under selection have all been extensively studied by Fisher (1930), Sewall Wright (1951) and others.

Finally, transcending in importance all problems of technique, is the choice of the objective of a breeding project. More breeding work has failed for lack of a clear objective than has stood still for lack of suitable genetic material for improvement. A general statement such as 'higher yield' or 'better quality' is not enough. The objective must be precisely defined in terms of the morphology and physiology of the plant if there is to be any precision in the estimation of the rate of advance. It is for this reason, and not because the problems involved are easier than others facing the plant breeder, that major successes have been achieved in cotton breeding in the synthesis of resistance to the jassid pest and bacterial blight disease. Progress sprang from the precise assessment of the character in which improvement was desired, and it is no accident that the best example of steady progress in the improvement of yield is Manning's (1956 b) study of yield advance in BP 52, where the components of yield have been defined and precisely measured from the beginning of the project.

6

THE IMPROVEMENT OF
AFRICAN COTTONS

Though it appears that the cultivated cottons arose from a wild African cotton (Hutchinson, 1954), there was a long interval between the domestication of *G. herbaceum* race *africanum* and the establishment of cotton in Africa as a cultivated plant. *G. arboreum* was differentiated from *G. herbaceum* and the *arboreum* cottons were separated into the northern group and the *indicum* group (Hutchinson, 1954) before the first crop of cotton was grown in Africa. The establishment of *G. arboreum* race *soudanense* in the agriculture of the Meroitic civilisation on the great bend of the Nile, its spread westward to West Africa, and the subsequent introduction and spread of *G. herbaceum* race *acerifolium* by the Moslem invaders, have been discussed elsewhere (Hutchinson, 1949).

These diploid cottons are now no more than relics of past cultures, having given place to the allopolyploid cottons of the New World. The more primitive of the New World allopolyploid cottons were introduced into Africa in the early period of transatlantic traffic, largely associated with the slave trade. The kidney cotton, *G. barbadense* race *brasiliense*, is still to be found wherever the slavers, European or Arab, established their routes. The *vitifolium* and *peruvianum* forms* of *G. barbadense* were established in the region of the Niger delta and spread inland into the forest and orchard bush regions respectively of what is now Nigeria.

G. hirsutum race *punctatum* was introduced on the Senegal coast at about the end of the eighteenth century (Hutchinson,

---

* Though these have in the past been accorded species rank, they are no more than agricultural forms, and were it not for the difference in their ecological adaptatation they would warrant no recognition.

1949) and spread rapidly across the savannah regions between the Sahara and the equatorial forests, covering the greater part of the tract formerly occupied by the two diploid species. About half a century later, the West Indian *G. hirsutum* race *marie-galante* was established in the Gold Coast by adherents of the Basle Mission (Hutchinson, 1949).

There was thus established a great range of New World allopolyploid material on which to base fresh cotton cultures in Africa. These introductions, however, all came from the more primitive cotton cultures of the New World, and the agriculture of Africa, apart from the Nile Valley, was backward in the extreme. Thus all the introductions were perennials, and perennials were, until the beginning of the present century, adequate for the needs of the rain-fed regions of the continent.

The first modern cotton crop to be established on the African continent began with Jumel's cotton in Egypt. The history of the venture has been set out by Dudgeon (1917) and from his record it is evident that Jumel's original perennial, brought by Maho Bey from the Sudan, was derived from the *vitifolium* stock which was established in the Nigerian forest belt, and spread thence eastwards to the Sudan along the trade routes.

The success of the new crop was such as to stimulate the reorganisation of the old basin irrigation from the Nile flood. Canals were deepened to allow of a flow at low river, and water lifts installed to raise the water to the level of the fields in which the perennial cottons were grown. Soon more ambitious projects were undertaken, barrages were built and the high-level perennial flow canal system was begun.

The demand for perennial irrigation arose from the perennial habit of the *vitifolium* cotton, and was, of course, made financially possible by the high value of the crop in the export market. Its success was such, however, as to encourage the introduction of a great range of cotton varieties from other countries. Most of these failed, but the annual Sea Islands had a limited success, and were grown on a small scale for some years. The Sea Islands belong to *G. barbadense*, the species

5-2

which also includes the established perennial *vitifoliums*. The two hybridised freely giving rise to a fertile vigorous segregating population, out of which high-quality annual types were selected, the forerunners of the modern Egyptian cottons. Thus the transformation of the Egyptian irrigation system, undertaken to meet the needs of a perennial cotton crop, was barely complete by the time the perennial cottons had been supplanted by a new race of annuals. Moreover, though the species to which they belong is of American origin, and was unknown in Africa until the sixteenth century at the earliest, the Egyptian cottons can truly be said to be endemic in the Nile Valley. They are in fact so closely adapted to the conditions under which they were developed that, apart from the Aden crop, a small crop in irrigated valleys in north Peru, and a small and economically rather precarious crop under irrigation in Arizona, Egyptian cotton has not been established successfully outside the Nile Valley.

The recent history of the Egyptian cottons provides a good illustration of the process of adaptation of a new crop to its environment. In Egypt new varieties have succeeded each other in rapid succession. Once the perennial irrigation system was established, the development of annual cottons made possible a much more intensive use of the land. Intensive land use, together with the ravages of pink bollworm on the slow-cropping long-term annuals that were first used, put a high premium on earliness, and earlier better-cropping varieties came out of the research farm at Giza in rapid succession. By contrast, in the Sudan with its very high summer temperatures and the strict limitation on the use of water for irrigation at low river, double cropping of the land is impossible. Under these conditions the crop has developed on entirely different lines from those followed in Egypt. Shorter term cottons have little advantage and the old Egyptian Domains Sakel variety is still cultivated twenty years after it disappeared from Egypt. In the Sudan two diseases, the virus disease known as leaf-curl and the bacterial blight commonly called blackarm, developed in epidemic form. The threat to the crop was in both cases met in the first instance by strict plant sanitation measures. Uprooting and burning of all cotton stalks was rigorously enforced. The rotation was altered and after cotton the land was left dry and fallowed, so that ratoons that might carry over leaf-curl to the new crop could be seen and uprooted, and were in any case starved of water. And leaf debris that might carry bacterial blight was swept up and burnt over the whole of the 200,000 acres of cotton land in the main irrigated area of the Sudan, the Gezira.

Under these circumstances, the advantages of genetic resistance both to leaf-curl and to blackarm were very considerable. Stocks resistant to leaf-curl were isolated and multiplied by Lambert (1938), and his 1730 strain became the standard variety for the southern half of the Gezira, in which the risk of leaf-curl was greatest. Resistance to bacterial blight was more difficult to achieve, since the gene content of the Egyptian cottons did not appear to provide material from which to build it up. Knight (1954) found a number of genes for resistance in cottons belonging to *G. hirsutum*, the first being identified in Uganda Uplands. His gene transference work is the classic in that field of applied genetics, and has resulted in the production of a range of bacterial blight-resistant Egyptian cottons closely matching in quality characteristics the two accepted commercial cottons of the Sudan.

The very different circumstances of Egypt and the Sudan, and the different breeding policies to which they gave rise, have resulted in a considerable divergence between the varieties of the two countries. Whereas in the past the Sudan seed supply was successfully based on imports of Domains Sakel seed from Egypt, modern Egyptian varieties such as Karnak are worthless in the Sudan, and modern Sudanese varieties such as the 1730 range would be quite out of place in Egypt.

Egyptian cotton provides the bulk of the high-quality cotton of the world's markets. In Egypt, the gains that have been made in agricultural characteristics have been such that within the accepted quality range of the country, the fate of a new variety has depended almost entirely on its yield. There have therefore been from time to time wide changes in the quality structure of the crop, and spinners have had to adjust their production programmes accordingly. In the Sudan, on the other hand, it was established as a matter of policy early in the history of irrigated cotton growing that varietal changes involving a change in quality would be avoided as far as possible. There is no doubt that this continuity in quality gave the Sudan a market reputation that resulted in a degree of independence of the fluctuations of the much larger Egyptian crop. Some sacrifice was involved, however. Not only was the promising NT 2 variety kept out of cultivation because it did not fit into the current quality structure of the crop, but the plant breeders were given very narrow quality limits within which to breed. The breeding policies of Egypt and the Sudan in regard to quality were thus exactly opposite. That pursued in Egypt resulted in considerable inconvenience to the country's customers. That in the Sudan somewhat handicapped the plant breeders, and probably resulted in a reduction of the return obtained from their work. With a proper balance between work of long-range interest and that offering an immediate return, it should be possible to breed fairly well to quality standards, and the wide fluctuations characteristic of the Egyptian crop in the past should be avoidable. On the other hand, some change in quality standards may be in the best interests of both producer and consumer, and it is now accepted that the quality limits imposed on plant breeders in the Sudan in the past may be relaxed.

The development of crops of Upland (*G. hirsutum* race *latifolium*) cotton in the rain-fed regions of Africa did not begin until the first decade of the present century, almost ninety years after the beginning of Egyptian cotton cultivation. In British Commonwealth territories, the basic stock used came from importations of the long-staple Upland cottons grown at that time in the United States Cotton Belt. Little was known in Africa of the technique of maintaining seed purity, and indeed cross-contamination between varieties went on to such an extent that varietal names are of no more than historical interest. The whole of the introduced material, which was derived from a few of the better quality American stocks, very shortly became a mixed and interbreeding bulk.

Distinctive types arose in most cotton-growing territories. One of the most successful of these was developed on European farms in Nyasaland and was widely distributed under the name of Nyasaland Upland on the eastern side of the African continent as far north as the Sudan. Selection work was started in Uganda about 1916, at Barberton in the Union of South Africa in 1924, and in other territories about the same time as at Barberton.

The jassid-resistant U 4 stocks bred by Parnell and his colleagues (Parnell, 1935) at Barberton constituted virtually a new race of Upland cotton. Since jassid is a damaging pest throughout South, Central and East Africa, and in the rain-fed regions of the Sudan, U 4 derivatives were widely distributed and contributed to some extent to the breeding stocks throughout the territories on the eastern side of Africa. The crops of South Africa, Portuguese East Africa, Southern Rhodesia and Nyasaland are all descended, with doubtless some admixture from earlier local stocks, from Barberton U 4. A U 4 stock was distributed in a part of the Lake Province of Tanganyika, and though later replaced by a local selection, contributed to the gene pool from which more recent improved strains have been derived. In Uganda, U 4 was grown on the Serere Experiment Station, and the Serere strain SP 84 was selected in a U 4 stock that was probably contaminated by natural crossing with local Uganda material. Though SP 84 was never grown extensively in Uganda, it provided the basic stock from which the varieties now grown in the rain-fed areas of the Sudan were bred.

In Tanganyika it became evident when breeding work was extended about 1940 that jassid resistance was of great importance in the Lake Province but of less value in the Eastern Province. A series of high-yielding jassid-resistant stocks has been distributed in the Lake Province from the Ukiriguru station under the designation Uk followed by a number indicating the year of issue. The length and fineness characteristic of the Lake Province crop has been maintained, but any improvement in either character has been deliberately avoided. Nevertheless, the great agronomic improvement that has followed the achievement of resistance to jassid has been accompanied by a very marked improvement of the market reputation of the crop. There can be little doubt that this quality improvement is a direct reflection of the better growth of the current jassid-resistant strain, and the consequent improvement in the nourishment of the lint.

The elimination of jassid as a serious pest has made more evident the damage done by bacterial blight, and the establishment of blight-resistant stocks is now the objective of the Ukiriguru plant breeders. Thus the improvement of the Lake Province cottons provides an excellent example of planned breeding work, in which the objectives at each stage are clearly defined and are attained before new objectives are added.

In Uganda the easy climate and the nature of the stocks available made possible the selection and establishment of varieties of higher quality than have been achieved in any other rain-fed region of Africa. The present Uganda crop is descended from three stocks selected at the Bukalasa Experiment Station, BP 52, BP 50 and B 181. BP 52 is a high-quality type in the same quality category as Egyptian Ashmouni. It is grown in the regions with a very equable climate west of the Nile. A derivative of BP 50, known as S 47, which is of slightly lower quality, but is rather more jassid resistant, is grown east and north of the Nile. B 181, which is a rather longer term variety inferior to the other two except at the highest yield levels, is not in general cultivation but is maintained as a source of resistance to *Verticillium* wilt, and has provided promising stocks in hybrids with BP 50.

Improvement of African Cottons

In commercial cotton stocks in Uganda, the rate of deterioration in general cultivation is high. All high-yielding varieties so far produced have fuzzy seeds. Naked, or nearly naked, seed is associated with low-ginning out-turn (Faulkner, in Namulonge P.R., 1956), and for reasons that are not yet understood, these low-ginning 'black seed' types increase at the expense of fuzzy seeded types. Precise assessment of the loss of crop consequent on this type of deterioration is very difficult, though some indication has been given by Walton (1957). Current breeding work (Manning, 1956b) provides annually a nucleus stock of pedigree BP 52 seed to form the basis of a continuous replacement system whereby the stock may be maintained at a high level of purity throughout the country.

The diversity that remains in BP 52 is still sufficient to allow of further improvement under Manning's selection index breeding technique. It is to be expected, however, that the variability will at some time be exhausted, and then further progress will depend on the exploitation of other, more variable, stocks. As a counterpart to the BP 52 breeding work, therefore, stocks have been set up to provide variable alternatives when the potentialities of the BP 52 material are exhausted by increasing homozygosity.

Two stocks have been established, one for fairly rapid exploitation in the near future, and one as a long-term asset. The more immediately useful stock, known as Albar, was established as a result of the study (Knight & Hutchinson, 1950) of the origin and distribution of resistance to bacterial blight in New World cottons. Since the *punctatum* cottons acclimatised in West Africa include a proportion of plants that are virtually immune to the disease, and since the Upland crop of northern Nigeria (Nigerian Allen) has been exposed to introgression from *punctatum* for many years, Knight and Hutchinson predicted that a proportion of Upland plants virtually immune to bacterial blight might be found in commercial stocks of Nigerian Allen. The prediction was verified at Shambat in the Sudan by the isolation in 1949 of a single almost immune plant from a small plot of Nigerian Allen grown from commercial seed from Funtua ginnery. This gave rise to the stock known as Albar 49. Subsequently a much larger test was carried out at Namulonge in Uganda. Funtua ginnery seed was again used. Six thousand seeds were heavily infected with bacteria before sowing. All seedlings showing infection were removed and the survivors sprayed with a suspension of bacteria. Again all plants that succumbed to the disease were removed, and the survivors again sprayed. A final elimination of infected plants left thirty-nine plants that had survived all attempts to infect them with the disease (Wickens, in Namulonge P.R., 1951). These thirty-nine plants gave rise to the Albar 51 stock. Subsequent breeding work has been devoted to the establishment of a very high degree of blight resistance in a genetically homozygous form, and to the improvement of yield and quality. In early generations, wide segregation occurred, and much material giving evidence of its *punctatum* ancestry was discarded. Albar material has been used as a source of high resistance to blight in breeding projects at a number of other African stations besides Namulonge, and besides offering the prospect of good protection against bacterial blight it has been shown by Walton (Uganda P.R., 1957) to possess valuable yield and quality characteristics. Recent improvements in the world's cottons (cf. the evolution of the Egyptian cottons discussed above) have been made most frequently where there has been a release of genetic variance following hybridisation between two races of the same species of *Gossypium*. The Albar stock provides breeding material of the type that has in the past been the foundation of extensive advances in the commercial cottons.

A more fundamental approach to the problem of providing a highly variable gene pool for plant-breeding improvement arose from consideration of the source of the East African commercial cottons in relation to the diversity available in the group to which they belong. The introductions from which almost the whole of the present East African stock is descended consisted of not more than half a dozen long-staple commercial Upland cottons, grown in the south-eastern States of the United States before the invasion of the Cotton Belt by boll weevil. The race *latifolium* of *G. hirsutum* includes a vast range of material not to be found in these limited stocks. Some of them, notably the modern varieties of the United States Cotton Belt and the West African Uplands, are known to contain gene material of value to East Africa. Others may well carry genes that would be of value either now or in the future in combination with East African stocks. It was therefore considered worth while to list the major *latifolium* stocks, and to select a number for inclusion in a breeding project aimed in the first instance at the synthesis of a highly variable population in which free gene exchange would take place. In such a population one may predict that a great diversity of genotypes will occur, among which selection may profitably be exercised.

The major types of *latifolium*, together with the stocks used and the more important characters desired from them, are listed in Table 3:

Table 3

Type	Stock	Characters desired
American Upland	Acala	Big boll, high yield
	Stoneville 20	Bacterial blight resistance
Guatemala *latifolium*	Four stocks from peasant crops	Persistent cropping
Chiapas *latifolium*	Acala district stock	Big boll
Oaxaca *latifolium*	Three stocks from peasant crops	Dense leaf hair
African Upland	U 4 (two strains)	Yield and jassid resistance
	Albar 49	Bacterial blight resistance
	BP 52	Quality
Indian Cambodia	Three Coimbatore strains and MU 8	Yield and jassid resistance
Indian Upland	Punjab LSS	Yield and bacterial blight resistance

To build up a population in which the gene content of these nineteen stocks is reassorted at random is evidently a long-term project, and any attempt at a systematic synthesis would be very time-consuming indeed. While there can be no substitute for time in the sense that recombination only occurs once in each generation, it is possible to minimise the time spent on the project by the plant breeder, by leaving the recombination to nature. The seed of the nineteen types was mixed and sown as a bulk. At flowering time interbreeding was ensured by transferring pollen from flower to flower with a camel-hair brush. The interpollinated flowers were marked, and the resulting seed cotton harvested as a bulk. The process is repeated annually. Since it must inevitably take many generations for the compete reassortment of the genotypes, it is important that the panmixis population thus established shall be maintained long enough for its potentialities to be realised. This is not to say that good material that may emerge from time to time cannot be withdrawn for use, but only that any such withdrawal must be arranged in such a way as to avoid the risk of skimming the cream from the continuing panmixis population. With cotton this is easily done. Not all bolls on a plant are interpollinated, and any withdrawals are made by taking naturally pollinated bolls only. Thus the genotype of the superior plants selected for more intensive breeding still contributes in full to the panmixis population.

In three generations the distinguishing features of the original strains completely disappeared, and a highly variable population arose containing material that it seemed worth while to withdraw and subject to selection. Single plant selections were made, but it was decided that they should not be inbred in progeny rows. Their seed was mixed to make a panmixis selection bulk. All plants were interpollinated, and again superior plants were selected and their seed bulked. Thus a lower intensity of selection was practised—virtually a selection based on the maternal half of the genotype only—and interbreeding among the selections was maintained. Though it is as yet too early to assess the value of the technique, it is already apparent that a highly variable population is easily synthesised in this manner, and that the characteristics of such a population are susceptible to change under the modest degree of selection that has been practised in the panmixis selection lines. Thus, as a

Improvement of African Cottons

long-term plant-breeding reserve, maintained at very little cost, a bulk of diverse origin maintained under a system of inter-pollination offers attractive possibilities.

In planning his breeding work with BP 52, Manning (1949) attempted some definition of the environmental circumstances for which he was breeding. The climatological study then initiated led to an assessment of the relation between planting data and crop yield, and thence on the one hand to a study of the rainfall regime of the Uganda cotton growing areas (Manning, 1956a) and on the other to studies of the water requirement of the cotton crop, and the extent to which crop size is limited by the rainfall regime (Manning & Farbrother in Namulonge P.R., 1956; Hutchinson, Manning & Farbrother, 1958).

The success with which the relation between sowing date and crop production was elucidated encouraged an attempt to go further and prepare a physiological specification for a cotton that would have a high yielding capacity in the area of Uganda in which BP 52 is grown. Climatic and crop-production studies at Namulonge (Manning & Farbrother, in Namulonge P.R., 1956) indicated that when rainfall was adequate the best yields were obtained from plantings made in the second half of June. Rainfall distribution studies, on the other hand, showed that the chance of receiving adequate rainfall for planting at that time is lower than at any other time during the period May to September in which the cotton crop must be established. Attempts have been made from time to time to breed an early cotton to be planted later, when more reliable planting conditions are experienced. These have never been successful, and it appears from water-requirement studies (Hutchinson, Manning & Farbrother, 1958) that an important reason is that if planting is postponed to late July or August the chance of the crop receiving adequate rainfall to supply its needs at peak leaf area is small.

There remains the alternative of putting forward the planting date to the beginning of June. While it is known that with existing varieties early June plantings generally yield a little

77

less than plantings in the second half of June, there is reason to believe that a variety might be bred that would give high returns from early June plantings. It was long thought that the cause of the lower yields of late May plantings was the more severe attacks of *Lygus* experienced by the earlier planted crops. Geering & McKinlay (in Namulonge P.R., 1953) have shown, however, that with early-planted cotton, recovery from early *Lygus* attack may be complete. Moreover, in some of their experiments they were able to show that crops subjected to an early attack of *Lygus* might even give higher yields that crops protected from the insect throughout their life. This finding was interpreted as evidence that a slight postponement of cropping may be advantageous in an early-planted crop. Thus, the breeding of a rather later fruiting cotton to be planted earlier might well be a profitable project.

This view received considerable further support from physiological studies. Farbrother (in Namulonge P.R., 1953) has demonstrated a relation between leaf area and crop size, and also (in Namulonge P.R., 1956) between leaf area and crop water requirement throughout the life of the plant. He has shown that under Namulonge conditions the crop is limited by water strain when the leaf area is high. Hence the crop may be said to be determined by the water available at or near the peak of leaf area. Thus any attempt to increase crop size by breeding for a plant with a greater peak leaf area would be likely to fail from inadequate moisture at the peak of crop development. Farbrother (in Namulonge P.R., 1955) has also shown, however, that crop development proceeds at a rate in excess of the rate of increase in total dry matter, and that it may be inferred that reserves are mobilised and transferred to the bolls from other parts of the plant. It should then be possible to increase crop production by increasing the reserves available for mobilisation at the peak of demand. Thus, by breeding a type with a longer vegetative period, reserves might be built up while leaf area was not limited by the moisture available, for use when the leaf area/soil moisture relationship becomes critical.

On the basis of these speculations a specification may be set out for a cotton plant to give the maximum yield under the climatic conditions of the BP 52 area of Uganda. The commercial Upland cottons are early annual derivatives of a genus of perennial shrubs. The development of the annual habit has gone on by the progressive reduction of the vegetative structure of the plant until little is left but a lax inflorescence mounted almost direct on the root system. By reversing the direction of selection of the last two or three centuries, it should be possible to re-establish a part of the old vegetative structure, and so to lengthen the period from sowing to flowering, thereby making possible sowing under the more favourable moisture conditions in May without advancing the picking season into the end of the second rains in October or November, and by thus increasing the leaf area during the vegetative phase to establish greater reserves in the plant before the peak of crop development, and thereby to get a bigger crop set without intensifying the water strain at the most difficult period.

Not only has physiological and climatological work made it possible to specify a type of cotton fitted to a particular environment, but the policy of setting up breeding stocks for long-term exploitation has provided material in which to select the types required. In the *latifolium* panmixis, and in no other stock that was available, was found the diversity in vegetative development in which selection for a longer term cotton variety could be made.

7

CONCLUSION

It is in the very nature of this account that it is an unfinished story. The Cotton Research Station was founded in Trinidad in 1926 in the belief that the acquisition of fundamental knowledge of the genus *Gossypium* would lead to advance in the breeding and growing of cotton in British Commonwealth countries. The Trinidad station was closed in 1944 and the Barberton station handed over to the Government of the Union of South Africa in 1949 in recognition of the fact that the time had come to bring together fundamental knowledge and practical experience in one of the more important of the African cotton growing territories. Seven years' experience at Namulonge in Uganda has provided ample evidence of the fruitfulness of the union, but the partnership is still in its infancy.

Cotton, which came out of Africa in the beginning, has only very recently become established as one of the economic pillars of the continent. It is in a state of rapid development in response to the demands and opportunities of African agriculture, and it offers enormous scope for the imaginative plant breeder. Those who work on cotton in Africa have a hand in crop-plant evolution. They can try out the theories of population genetics and the techniques of plant breeding with the expectation of seeing within their own lives how they work.

I have attempted to set out something of the ideas and methods that have grown from the elucidation of species relationships in Trinidad and the first application of statistics to cotton breeding in Indore. The latter part of those stories is bound up with the beginnings of studies of the deliberate management of variability in plant breeding, and of the enlistment of the plant physiologist to define the environment and to specify the plant breeder's problem. Thus the work of Namulonge is no mere exploitation of the combination of the theories of Trinidad and Indore with the experience of Barberton. In the developing partnership between geneticists and physiologists lies the hope of understanding the nature of our crop plants, and of directing the improvement of crop production.

REFERENCES

BARBOSA, D., trans. by STANLEY, H. E. J. (1866). *Hakluyt Soc.* **5**, 13.
BEASLEY, J. O. (1942). *Genetics*, **27**, 25.
BIRD, J. B. (1948*a*). *Soc. for Amer. Archaeol.* Mem. **4**, 21.
BIRD, J. B. (1948*b*). *Natural History*, September 1948, p. 296.
BIRD, J. B. & MAHLER, J. (1951–2). *American Fabrics*, no. 20, 73.
BRINKERHOFF, L. A., GREEN, J. M., HUNTER, R. & FINK, G. (1952). *Phytopathology*, **42**, 98.
CHEVALIER, A. (1935). *Rev. Bot. Appl.* **13**, 190.
DOUWES, H. (1953). *J. Genet.* **51**, 611.
DUDGEON, G. C. (1917). *Min. Agric. Egypt*, 1916, no. 3*a*.
EDLIN, H. L. (1935). *New Phytol.* **34**, 1 and 122.
EXELL, A. W. & HILLCOAT, D. (1954). *Contr. para o Conhecimento da Flora de Moçambique*—II, 55. Ministerio do Ultramar, Lisboa.
EXELL, A. W. & MENDONCA, F. A. (1954). *Contr. para o Conhecimento da Flora de Moçambique*—II, 63. Ministerio do Ultramar, Lisboa.
FISHER, R. A. (1930). *The Genetical Theory of Natural Selection.* Oxford Univ. Press.
GENTRY, H. S. (1956). *Madroño*, **13**, 261.
GERSTEL, D. U. (1953). *Evolution*, **7**, 234.
GERSTEL, D. U. (1956). *Genetics*, **41**, 31.
GERSTEL, D. U. & SARVELLA, P. A. (1956). *Evolution*, **10**, 408.
GULATI, A. M. & TURNER, A. J. (1928). *Ind. Cent. Cotton Cttee.* Tech. Lab. Bull. 17.
HARLAND, S. C. (1933). *C.R. Acad. Sci. U.S.S.R.* **4**, 181.
HARLAND, S. C. (1939). *The Genetics of Cotton.* Jonathan Cape.
HARLAND, S. C. (1950–1). *Mem. Proc. Manc. Lit. Phil. Soc.* 1.
HOWARD, A. & HOWARD, G. L. C. (1909). *Mem. Dept. Agric. Ind.* (Bot. Ser.), II, 7.
HUTCHINSON, J. B. (1940). *J. Genet.* **40**, 271.
HUTCHINSON, J. B. (1944). *J. Agric. Univ. Puerto Rico*, **28**, 35.
HUTCHINSON, J. B. (1947). *New Phytol.* **46**, 123.
HUTCHINSON, J. B. (1949). *Emp. Cott. Gr. Rev.* **26**, 1.
HUTCHINSON, J. B. (1951). *Heredity*, **5**, 161.
HUTCHINSON, J. B. (1954). *Heredity*, **8**, 225.
HUTCHINSON, J. B., GADKARI, P. D. & ANSARI, M. A. A. (1938). *Proc. 1st Conf. Sci. Res. Workers on Cotton in India*, p. 296. I.C.C.C. Bombay.

References

HUTCHINSON, J. B., KNIGHT, R. L. & PEARSON, E. O. (1950). *J. Genet.* **50**, 101.

HUTCHINSON, J. B. & KUBERSINGH (1936). *Indian J. Agric. Sci.* **6**, 672.

HUTCHINSON, J. B. & LEE, B. S. M. (1958). *Kew Bull.* no. 2.

HUTCHINSON, J. B. & MANNING, H. L. (1943). *Emp. J. Exp. Agric.* **11**, 140.

HUTCHINSON, J. B. & MANNING, H. L. (1945). *Emp. J. Exp. Agric.* **13**, 80.

HUTCHINSON, J. B., MANNING, H. L. & FARBROTHER, H. L. (1958). *J. Agric. Sci.* **51**, 177.

HUTCHINSON, J. B. & PANSE, V. G. (1937). *Indian J. Agric. Sci.* **7**, 531.

HUTCHINSON, J. B., SILOW, R. A. & STEPHENS, S. G. (1947). *The Evolution of Gossypium*. Oxford Univ. Press.

HUTCHINSON, J. B. & STEPHENS, S. G. (1944). *Trop. Agriculture, Trin.* **21**, 123.

JAMESON, J. D. & THOMAS, D. G. (1952). *Prog. Rep. from Exp. Sta. Uganda*. E.C.G.C. London.

JOHNSON, F. (1951). *Soc. Amer. Archeol.* Mem. 8.

KERR, T. (1951). *Proc. Mtg. Southern Agric. Workers*, Memphis, Tenn., U.S.A.

KNIGHT, R. L. (1952). *J. Genet.* **51**, 47.

KNIGHT, R. L. (1954). *Emp. J. Exp. Agric.* **22**, 68.

KNIGHT, R. L. & HUTCHINSON, J. B. (1950). *J. Genet.* **50**, 36.

LAMBERT, A. R. (1938). *Emp. Cott. Gr. Rev.* **15**, 14.

MACDONALD, D., FIELDING, W. L. & RUSTON, D. F. (1940). *Prog. Rep. from Exp. Sta. Barberton*. E.C.G.C. London.

MANNING, H. L. (1949). *Emp. J. Exp. Agric.* **17**, 245.

MANNING, H. L. (1956*a*). *Proc. Roy. Soc.* B, **144**, 460.

MANNING, H. L. (1956*b*). *Heredity*, **10**, 303.

MASON, T. G. (1938). *Emp. Cott. Gr. Rev.* **15**, 113.

MAUER, F. M. (1930). *Bull. Appl. Bot. Pl. Breed.* Suppl. to the Suppl. **47**, 425.

MUDALIAR, V. R. & BALASUBRAMANYAN, R. (1951). *Indian Cott. Gr. Rev.* **5**, 176.

NAMULONGE P.R. (1951, 1953, 1955, 1956). Annual *Progress Reports of the Cotton Research Station, Namulonge*. E.C.G.C. London.

PARNELL, F. R. (1925). *Prog. Rep. from Exp. Sta. Barberton*. E.C.G.C. London.

PARNELL, F. R. (1935). *Emp. Cott. Gr. Rev.* **12**, 177.

PARNELL, F. R., KING, H. E. & RUSTON, D. F. (1949). *Bull. Ent. Res.* **39**, 539.

References

RAMANATHA AYYAR, V. (1938). *Proc. 1st Conf. Sci. Workers on Cotton in India*, p. 365. I.C.C.C. Bombay.

RICK, C. M. (1950). *Evolution*, 4, 110.

SILOW, R. A. (1941). *J. Genet.* 42, 259.

SILOW, R. A. (1944). *J. Genet.* 46, 62.

SKOVSTED, A. (1937). *J. Genet.* 34, 97.

SMITH, H. FAIRFIELD (1936). *Ann. Eugen.* 7, 240.

SMITH, J. (1950). *Distribution of Tree Species in the Sudan in Relation to Rainfall and Soil Texture.* McCorquodale, London.

STEBBINS, G. L. (1947). *Ecol. Monogr.* 17, 149.

STEPHENS, S. G. (1946). *J. Genet.* 47, 150.

TOTHILL, J. D. (1940). *Agriculture in Uganda.* Oxford Univ. Press.

WALTON, P. D. (1957). *Emp. Cott. Gr. Rev.* 34, 99.

WATT, SIR G. (1907). *The Wild and Cultivated Cotton Plants of the World.* Longmans, London.

WEINDLING, R. (1948). *U.S.D.A. Tech. Bull.* 956.

WHITAKER, T. W. & BIRD, J. B. (1949). *American Museum Novitates*, no. 1426. New York.

WRIGHT, SEWALL, (1951). *Ann. Eugen.* 15, 323.

UGANDA P.R. (1957). *Prog. Rep. from Exp. Sta. Uganda.* E.C.G.C. London.

INDEX

Index

Index